AF893236

# INTRODUCTION TO CONVOLUTIONAL CODES WITH APPLICATIONS

# THE KLUWER INTERNATIONAL SERIES IN ENGINEERING AND COMPUTER SCIENCE

## COMMUNICATIONS AND INFORMATION THEORY

*Consulting Editor*

**Robert Gallager**

***Other books in the series:***

**CODED-MODULATION TECHNIQUES FOR FADING CHANNELS,** S. Hamidreza Jamali, and Tho Le-Ngoc
ISBN: 0-7923-9421-6
**ELLIPTIC CURVE PUBLIC KEY CYRPTOSYSTEMS**, Alfred Menezes
ISBN: 0-7923-9368-6
**SATELLITE COMMUNICATIONS: Mobile and Fixed Services**, Michael Miller, Branka Vucetic and Les Berry
ISBN: 0-7923-9333-3
**WIRELESS COMMUNICATIONS: Future Directions**, Jack M. Holtzman and David J. Goodman
ISBN: 0-7923-9316-3
**DISCRETE-TIME MODELS FOR COMMUNICATION SYSTEMS INCLUDING ATM**, Herwig Bruneel and Byung G. Kim
ISBN: 0-7923-9292-2
**APPLICATIONS OF FINITE FIELDS**, Alfred J. Menezes, Ian F. Blake, XuHong Gao, Ronald C. Mullin, Scott A. Vanstone, Tomik Yaghoobian
ISBN: 0-7923-9282-5
**WIRELESS PERSONAL COMMUNICATIONS**, Martin J. Feuerstein, Theodore S. Rappaport
ISBN: 0-7923-9280-9
**SEQUENCE DETECTION FOR HIGH-DENSITY STORAGE CHANNEL**, Jaekyun Moon, L. Richard Carley
ISBN: 0-7923-9264-7
**DIGITAL SATELLITE COMMUNICATIONS SYSTEMS AND TECHNOLOGIES: Military and Civil Applications**, A. Nejat Ince
ISBN: 0-7923-9254-X
**IMAGE AND TEXT COMPRESSION**, James A. Storer
ISBN: 0-7923-9243-4
**VECTOR QUANTIZATION AND SIGNAL COMPRESSION**, Allen Gersho, Robert M. Gray
ISBN: 0-7923-9181-0
**THIRD GENERATION WIRELESS INFORMATION NETWORKS**, Sanjiv Nanda, David J. Goodman
ISBN: 0-7923-9128-3
**SOURCE AND CHANNEL CODING: An Algorithmic Approach**, John B. Anderson, Seshadri Mohan
ISBN: 0-7923-9210-8
**ADVANCES IN SPEECH CODING**, Bishnu Atal, Vladimir Cuperman, Allen Gersho
ISBN: 0-7923-9091-1
**SWITCHING AND TRAFFIC THEORY FOR INTEGRATED BROADBAND NETWORKS**, Joseph Y. Hui
ISBN: 0-7923-9061-X
**ADAPTIVE DATA COMPRESSION**, Ross N. Williams
ISBN: 0-7923-9085
**SOURCE CODING THEORY**, Robert M. Gray
ISBN: 0-7923-9048-2
**AN INTRODUCTION TO ERROR CORRECTING CODES WITH APPLICATIONS**, Scott A. Vanstone, Paul C. van Oorschot
ISBN: 0-7923-9017-2

# INTRODUCTION TO CONVOLUTIONAL CODES WITH APPLICATIONS

*by*

**Ajay Dholakia**
*North Carolina State University*

SPRINGER SCIENCE+BUSINESS MEDIA, LLC

ISBN 978-0-7923-9467-9 ISBN 978-1-4615-2712-1 (eBook)
DOI 10.1007/978-1-4615-2712-1

**Library of Congress Cataloging-in-Publication Data**

A C.I.P. Catalogue record for this book is available
from the Library of Congress.

**Copyright** © 1994 by Springer Science+Business Media New York
Originally published by Kluwer Academic Publishers in 1994

All rights reserved. No part of this publication may be reproduced, stored in a retrieval system or transmitted in any form or by any means, mechanical, photo-copying, recording, or otherwise, without the prior written permission of the publisher, Springer Science+Business Media, LLC.

*Printed on acid-free paper.*

*To*

*my mother Mrs. Archana Dholakia*
*and*
*my father Dr. Harshad Rai Dholakia*

# CONTENTS

# LIST OF FIGURES

# LIST OF TABLES

**Chapter 4**

**Chapter 5**



**Chapter 6**



**Chapter 8**



**Chapter 11**

# PREFACE

Coding theory is by now a mature field, with over four and a half decades of research, teaching, development, and usage devoted to it. I believe that convolutional coding, one of its sub-fields, has also matured enough to merit independent treatment. Over the course of my doctoral research, I felt the need for a book devoted entirely to convolutional coding that covered all the fundamentals and summarized most of the recent research in the field. This book is the result of my efforts to fill the void. Its objectives are to provide an introduction to the basic concepts of convolutional codes and encoders, to discuss various convolutional decoding techniques, and to present some of their common applications. Limitations of size have led to the exclusion of communication theory aspects such as modulation and detection, and of a high-level algebraic treatment. A first draft of the book has also been used as a textbook for a senior level computer science elective on coding theory and applications, with major emphasis on convolutional coding.

This book is organized in four parts and eleven chapters. Examples and exercises are included to enhance understanding. Computer programming assignments are also included in many chapters to allow the use of this book in a design course.

- Part I is introductory and includes Chapter 1. It introduces the problem of digital communication over noisy channels. Commonly used error control protocols, error-correcting and detecting codes, and channel models are summarized. Applications of convolutional coding are pointed out.
- Part II is devoted to definition, distance structure, and classification of convolutional codes and encoders. Chapter 2 is on their definitions and representations. Chapter 3 is on distance properties in which various distance measures are defined and algorithms for computing them are discussed. Chapter 4 is on classification of convolutional codes and encoders. Commonly used as well as new classes of codes and encoders are included.

- Part III describes convolutional decoding techniques. Chapters 5 through 8 are on Viterbi, sequential, majority-logic, and table look-up decoding, respectively. Each decoding scheme is described with the help of complete examples. Main decoding algorithms are given and variations are pointed out, and their performance is analyzed. Performance improvement by using soft decisions is also described.

- Part IV presents applications of convolutional coding in hybrid ARQ error control protocols and in lost packet recovery in high-speed networks. A new classification of hybrid-ARQ protocols based on fixed or variable amount of coding redundancy is suggested. This classification allows a natural way of studying the underlying concepts of hybrid-ARQ schemes and accommodates new research in this area. Chapter 9 is devoted to fixed-redundancy ARQ schemes; conventional ARQ and convolutional hybrid-ARQ schemes based on the modification of various convolutional decoding techniques are summarized. Chapter 10 describes the application of convolutional codes in variable-redundancy hybrid-ARQ error control schemes. Chapter 11 describes a novel application of convolutional codes for lost packet recovery in high-speed networks, and hopefully opens the doors for further research.

Appendix A is a brief introduction to the basic concepts of abstract and linear algebra. Appendix B summarizes elementary aspects of block codes, and is useful in understanding error-detecting schemes in Chapters 9 and 10. Appendices C and D give brief introductions to coded modulation systems and concatenated coding systems. Appendix E is a short, annotated guide to convolutional coding literature. It points to a collection of over 450 original research articles and books. The references are next, followed by the index.

Practicing communications, electronics, and networking engineers will be able to get a grasp of the fundamental concepts of convolutional coding and its applications. This book is suitable for use as a textbook in a one-term course on coding theory with emphasis on convolutional codes. Senior undergraduate and graduate students in electrical and computer engineering as well as computer science can use it as a textbook or additional reference. Researchers in the field of coding theory can use it as a quick reference and as a preparation for studying the literature.

*Raleigh, N.C.* *Ajay Dholakia*

# ACKNOWLEDGEMENTS

This book has drawn on the efforts of many individuals. It is my pleasure to acknowledge their help here.

I would like to thank my doctoral advisors Dr. D.L. Bitzer and Dr. M.A. Vouk for their continued support and encouragement. Their patience while I was juggling the tasks of writing my dissertation and this book allowed me the freedom to maintain my own rhythm. I would also like to thank them for the opportunity to teach an introductory course on coding theory and applications (CSC 495K) in Fall, 1993, which provided a concrete platform to develop this book.

I am grateful to the students in CSC 495K, who came from various disciplines - computer science, electrical engineering, and mathematics, for their enthusiasm in learning this material. I especially want to thank Dr. R.E. Hartwig for the discussions during and after class through-out the semester.

Dr. K. J. Hole carefully read a draft of the book, and furnished a thorough and extensive review. Dr. G.D. Forney, Jr. provided useful comments that have influenced the structure of the book. I sincerely thank both of them for their invaluable and experienced input.

I wish to thank Mr. H. Koorapaty for his participation in lengthy discussions many of which were on topics included in this book.

I acknowledge the Department of Computer Science at North Carolina State University for providing the facilities. I also thank Dr. R.K. Cavin III of the Department of Electrical and Computer Engineering at NCSU for his interest and support.

I am thankful to Mr. R.W. Holland, Jr. of Kluwer Academic Publishers for his continued involvement since the beginning of this project. I am also thankful to Ms. Suzanne Rumsey and Ms. Rose Luongo at Kluwer for all their help.

I am indebted to my family for always believing in me and for their interest and support through-out this endeavor. Finally, I thank my wife Dharani for patiently and repeatedly adjusting her schedule in making my first such venture as smooth as possible. I thank her also for proof-reading the manuscript, and for her help in creating the bibliography. Her comments as an outsider to the field have been extremely valuable. This book would not have been possible without her love and support.

*A. D.*

# PART I

# INTRODUCTION

# 1

# ERROR CONTROL CODING: AN OVERVIEW

Reliable transmission of information is one of the central topics in digital communications. While the advent of powerful digital computing and storage technologies has made the data traffic increasingly digital, the reality of inherently noisy communication channels persists and has thus made error control coding an important and necessary step in achieving reliable communication. Coding in this case refers to systematically introducing redundancy in a message prior to transmission and using this redundancy to recover the original message from the erroneously received one. There are two basic types of error control codes: block codes and convolutional codes. Convolutional codes, techniques for decoding them and their applications are the focus of this book. This introductory chapter begins with a brief historical background of the development of information theory, a field of study that established the utility of error control coding.

## 1.1 HISTORICAL BACKGROUND

A concrete first step in electrical communication was taken over 150 years ago, when the first model of an electromagnetic telegraph was built in 1834 by Gauss and Weber. Other models of telegraph were quickly developed, e.g., Cooke and Wheatstone telegraph in 1837 and Morse's telegraph in 1842 which lead to the installation of transatlantic telegraph cables in 1866. Telephones were the next electrical communication technology to emerge, followed by wireless radio communication. The first half of the twentieth century saw the development and commissioning of television. Microwave communication, satellite communication, and fiber optic communication have now brought us into the modern era of electrical communication.

Early efforts to explore fundamental limits on the rate and the accuracy of information transmission were made in the 20th century by Nyquist [344] and Hartley [186]. Hartley was first to propose a quantitative measure of information based on the use of logarithms. This branch of study came into its own in the work of Claude E. Shannon, starting in 1948 with his famous series of "A Mathematical Theory of Communication" papers [413, 416, 427]. In this fundamental and ground-breaking work, Shannon established the fact that noise on a communication channel limits the rate but not the accuracy of information transmission, and founded the field of *information theory.*

Shannon's work spawned three sub-fields of information theory: *source coding*, *channel coding*, and *rate-distortion theory.* Source coding is concerned with efficient representation of information generated by a source, and usually attempts to reduce the amount of redundancy in the source information so that as high an information transmission rate can be achieved as is possible without jeopardizing the fidelity of the original source output. Shannon proved that the amount of information can be measured by a metric called the *entropy* of the source, and that the maximum achievable information transmission rate is proportional to the entropy of the source. Rate-distortion theory extends the ideas of source coding by allowing a certain amount of distortion in the reconstruction of source information at the receiver. For instance, an analog source output cannot be accurately reconstructed if it is converted, for the purpose of transmission, to discrete form with finite symbols in the discrete alphabet.[1]

Channel coding is necessary for minimizing the error rate in the received symbols when transmitting information over a noisy channel. Shannon proved that given the characteristics of a communication channel such as its bandwidth, the signal power, and the noise level, a quantity called the *channel capacity* can be derived. The channel capacity becomes an effective design parameter. Shannon showed that as long as the rate of information transmission is kept below the channel capacity, arbitrarily high accuracy can be achieved in the received information by using some form of coding. Unfortunately, Shannon's work provided only a nonconstructive proof of this basic result, and much of the last four and a half decades of research effort has focussed on devising practical means of realizing these results.

[1] The reader may refer to [37] and [165] for further details on rate-distortion theory and source coding theory. General treatment of information theory can be found in [159] and [454].

When dealing with sequences of discrete symbols as the common representation of information, channel coding is achieved by using ***block codes*** or ***convolutional codes***. Block codes were invented by Hamming [181]. Convolutional codes were invented by Peter Elias in 1955 [129] and have been the rival channel codes to block codes ever since. It is said that there are two camps within the channel coding community: the block coding types and the convolutional coding types. As pointed out by Forney [146], the relations between these two camps have been very amicable, and have resulted in the exchange of a wealth of knowledge over the last four decades. Owing to the maturity of the field of coding theory, convolutional coding can be defined as an independent sub-field and can be discussed, applied, and taught on its own.

## 1.2 A DIGITAL COMMUNICATION SYSTEM

A block diagram of a general digital communication system is shown in Figure 1.1. Observe that we have grouped the information source and the source encoder to form a discrete data source, the modulation and demodulation functions with the waveform channel to form a discrete channel, and the source decoder and information sink to form a discrete data sink. This diagram is described below in further detail.

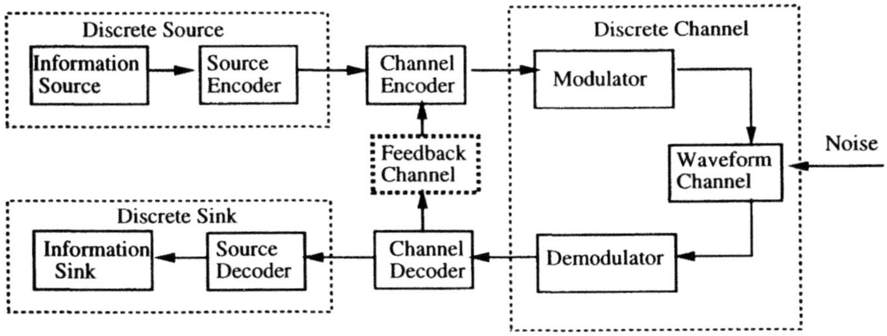


**Figure 1.1** A general digital communication system.

### 1.2.1 Discrete Data Source

The *discrete data source* produces a sequence of discrete symbols from a finite alphabet, and the discrete data sink expects a similar (preferably the same) sequence of discrete symbols as its input.

An *information source* is the originator of the message that has to be communicated to the intended user or information sink. The source can be analog, i.e., its output could be a continuous waveform, taking one of an infinite number of possible values at any time, or it can be digital, i.e., its output could be restricted to one of a finite number of values. Examples of an analog source are the telephone, the video camera, etc., and examples of a digital source is a computer sending a binary stream.

A *source encoder* acts by removing unnecessary redundancy in the source information without compromising the fidelity of the source. The objective is to represent the source information using as small a number of "bits" as possible in order to maximize the rate of information transmission. If the information source is analog, the source encoder also performs analog-to-digital conversion. This process is also known as quantization. Some information is lost in this process if a finite number of digital symbols are used to represent an analog signal. If the information source is digital, the source encoder may change the digital alphabet used to represent the information.

### 1.2.2 Channel Encoder

The *channel encoder* adds redundancy to the message to be transmitted in a well-defined manner. This is done to allow the decoder at the receiver site to correct errors that may have been caused by the transmission channel. A channel encoder may be a convolutional or a block encoder.

### 1.2.3 Discrete Channel

A *discrete channel* accepts discrete symbols as input and delivers discrete symbols as its output. It consists of a modulator, a waveform channel, and a demodulator.

A *modulator* converts a discrete symbol into a carrier waveform by assigning a different waveform to each possible symbol. Thus, there is a finite number of different waveforms corresponding to the finite number of possible symbols. For example, in *binary-phase-shift-keyed* (BPSK) modulation, a waveform $s_0(t)$ is assigned to a binary '0' and $s_1(t)$ to a binary '1', where

$$\begin{aligned} s_0(t) &= \sqrt{\tfrac{2E}{T}}\cos(w_0 t), \\ s_1(t) &= \sqrt{\tfrac{2E}{T}}\cos(w_0 t + \pi). \end{aligned} \tag{1.1}$$

Here, $E$ is the energy of each signal and $T$ is the period over which the signal is asserted. Note that $s_0(t) = -\ s_1(t)$.

A *waveform channel* is a continuous channel which allows transmission of an electrical signal of a certain carrier frequency over a physical medium or a succession of media. The output of such a channel is a waveform which may be distorted due to noise disturbances. Additive white Gaussian noise (AWGN) is a common type of noise disturbance on communication channels. In this case, let $n(t)$ be a Gaussian random process with one-sided *power spectral density* (PSD) $N_0$. Then the channel output $r(t)$ corresponding to the input $s(t)$ is

$$r(t) = s(t) + n(t). \tag{1.2}$$

The *demodulator* attempts to recover the transmitted waveform from the channel output $r(t)$. A matched filter followed by a sampler is typically used in an optimum demodulator. For BPSK modulation with coherent detection, the demodulator output is a real number $\rho$ given by

$$\rho = \int_0^T r(t)\sqrt{\frac{2E}{T}}\cos(w_0 t)dt. \tag{1.3}$$

The demodulator output can be directly passed on to the channel decoder in an unquantized form, in which case the decoder must be capable of handling analog input. In case of a discrete channel decoder, two possibilities exist. The demodulator output can be quantized into a finite number ($Q$) of levels. A 3-bit ($Q = 8$) quantization is commonly used. A demodulator with $Q > 2$ is called a *soft decision* demodulator. If the demodulator output is quantized into $Q = 2$ levels, it is called a *hard decision* demodulator.

For AWGN disturbance on the channel, the probability of error can be easily computed, and is given by

$$P_E = \frac{1}{\sqrt{2\pi}}\int_{\sqrt{2E/N_0}}^{\infty} e^{-x^2/2}dx. \tag{1.4}$$

It is possible to simplify this expression by writing it as

$$P_E = \frac{1}{2}\mathrm{erfc}(\sqrt{\mathrm{E}/\mathrm{N}_0}), \tag{1.5}$$

where

$$\mathrm{erfc}(\mathrm{x}) = \frac{1}{\sqrt{\pi}}\int_{\mathrm{x}}^{\infty} \mathrm{e}^{-\mathrm{y}^2}\mathrm{dy}. \tag{1.6}$$

The expression for the probability of error for other digital modulation schemes can be similarly obtained [450, 365].

### 1.2.4 Channel Decoder

The *channel decoder* accepts the demodulator output and generates an estimate of the original information message that was presented to the channel encoder. The channel decoder can be designed to work with either hard or soft decision demodulators. Typically, the use of soft decision demodulators results in improved performance since the soft quantized demodulator output can be treated as an indication of the reliability of the decoded symbol.

### 1.2.5 Discrete Data Sink

The *discrete data sink* is the intended "user" of the discrete data source output that was communicated. It may be a person, or an electrical or electronic device such as a loudspeaker or a computer. The *source decoder* attempts to reconstruct the original source information. If the information is to be delivered in a digital form, only an alphabet conversion may be involved. However, digital-to-analog conversion may be necessary if the information has to be delivered in an analog form. As noted earlier, some distortion has to be tolerated due to quantization.

### 1.2.6 Feedback Channel

An error control system based on error detection and retransmission requires a feedback channel, which is used by the receiver to send a positive or a negative acknowledgement to the transmitter in response to a message. The transmitter in turn responds by either sending the next message (if the acknowledgement was positive) or by retransmitting the previous message (if the acknowledgement was negative).

### 1.2.7 Digital Communication System Trade-Offs

The performance of a digital communication system is most often measured by the probability of error in the received symbol, i.e., by the probability that an error is made in estimating the symbol that was sent. The system designer's aim is to reduce this probability as much as possible while trying to increase the rate of information transmission and minimize implementation cost. The system resources available to the designer in achieving this goal are the signal power (or, equivalently, the signal energy per transmitted symbol) and the channel bandwidth. In a power-limited application, e.g., deep space communication, increasing the bandwidth by introducing coding may be possible. In a bandwidth-limited application, e.g., telephone and terrestrial radio, increasing signal power or expanding the signal constellation and introducing coding without bandwidth expansion are possible alternatives (see Appendix C).

Coding, i.e., adding redundancy to the transmitted information implies an increase in transmission bandwidth and complexity of (decoder) implementation. For a given rate of transmission, use of channel coding offers the following alternatives: the system bandwidth requirements can be reduced by increasing the signal power or relaxing the desired level of the probability of error, or the probability of error can be improved at the expense of signal power or channel bandwidth, or, finally, the signal power requirement can be reduced by increasing the channel bandwidth or relaxing the desired probability of error. Use of coding increases system complexity, but this increase is often worthwhile since coding can provide high accuracy with less energy than uncoded communication.

## 1.3 TYPES OF CHANNELS

In the study of digital communication systems, various models of the channels used are developed. While these models are not always realistic, they allow easy analysis of the system behavior and provide the system designer with valuable information regarding the trade-offs involved in various (admittedly idealized) scenarios. In some cases, the actual channel is accurately represented by a model, e.g., the deep space channel modeled by the additive white Gaussian noise channel. Examples of communication channels are telephone lines, cable, microwave, troposcatter, and HF radio, space (or satellite), and the recording channel.

### 1.3.1 Discrete Memoryless Channel

A common channel model is the *discrete memoryless channel* (DMC), shown in Figure 1.2(a), with an input alphabet consisting of $M$ symbols ($x_0$, $x_1$, $\ldots$, $x_{M-1}$) and an output alphabet consisting of $Q$ symbols ($y_0$, $y_1$, $\ldots$, $y_{Q-1}$). The memoryless property implies that the output at any time unit depends only on the input at that time and not on any previous input. The channel is specified by the transition probability $P(y_j|x_i)$, i.e., the probability of receiving the symbol $y_j$ given that the symbol $x_i$ was transmitted. A DMC is called *symmetric* if $P(y_j|x_i) = P(y_{Q-j-1}|x_i)$. A DMC with $M = 2$ is called a binary input $Q$-ary output DMC, and is often used to model a channel with a soft decision ($Q > 2$) demodulator.

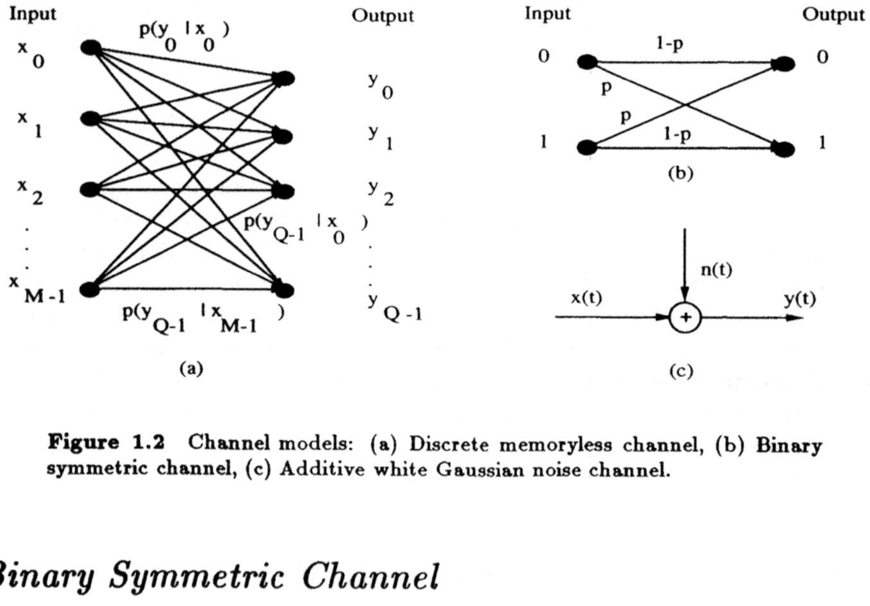


**Figure 1.2** Channel models: (a) Discrete memoryless channel, (b) Binary symmetric channel, (c) Additive white Gaussian noise channel.

## *Binary Symmetric Channel*

A simple and useful type of DMC is shown in Figure 1.2(b). The input as well as the output alphabet is $\{0, 1\}$, i.e., the DMC has $M = Q = 2$, and it is symmetric. A '0' (or '1') is transmitted in each time unit and is received correctly with a probability $(1 - p)$, where $p$ is the transition probability of the channel, given by

$$p = P(1|0) = P(0|1). \tag{1.7}$$

Thus, $p$ is the probability that the channel introduces an error. This channel is known as the *binary symmetric channel* (BSC). The BSC is used to model a discrete channel with a hard decision demodulator.

### *Additive White Gaussian Noise Channel*

A commonly used DMC is the *additive white Gaussian noise* (AWGN) channel in which the noise disturbance is assumed to be a zero-mean Gaussian random process with a one-sided spectral density $N_0$. A model of the AWGN channel is shown in Figure 1.2(c). Examples of AWGN channels include satellite and deep space communication channels. The output $y(t)$ of the AWGN channel is given by

$$y(t) = x(t) + n(t), \tag{1.8}$$

where $x(t)$ is the channel input and $n(t)$ is the white Gaussian noise process. Using this model, the conditional probability density function of the output $y(t)$ is given by

$$p(y|x = x_i) = \frac{1}{\sqrt{2\pi}\sigma} e^{-(y-x_i)^2/2\sigma^2}, \tag{1.9}$$

where $\sigma^2$ is the variance of the Gaussian noise.

## 1.3.2 Channels with Memory

Many communication channels used in practice are better described as channels with memory, e.g., burst HF channels, and magnetic and optical recording systems. The main characteristic of channels with memory is statistical dependence between errors. The errors tend to occur in large numbers, i.e., *bursts* which are interspersed with periods of essentially error-free communication. The error control strategies required to obtain performance improvements on channels with memory are quite different from those needed for memoryless channels. This book does not treat channels with memory any further. The reader may refer to [209, 123, 245, 249] for more information.

## 1.3.3 Nonstationary Channels

In many communication systems, the noise characteristics of the channel varies over a period of time. For instance, in a radio and satellite communication, multipath fading occurs due to shadows or clouds. This phenomenon is also experienced in mobile radio communication. This leads to fluctuations in the noise levels and any *a priori* estimate of the coding redundancy required to perform error control may be rendered futile and may result in severe performance degradation. Such channels are called *nonstationary.* Adaptive error control is an effective way to provide reliable communication over nonstationary channels.

## 1.4 ERRORS, CODES, AND PROTOCOLS

### 1.4.1 Types of Errors

The memoryless channels discussed in the previous section have the property that each transmitted symbol is affected by noise independent of the others. This results in a natural "randomization" of channel errors, and this type of errors are called *random errors*. Deep space channels and satellite channels suffer from random errors and are effectively modeled as memoryless channels.

Channels with memory have noise characteristics quite different from those of memoryless channels. A cluster of transmitted symbols on a channel with memory is affected by noise with a high probability, and then there are large "silent" (i.e., noise-free) periods. Thus, the channels seems to be in a "good" state for most of the time, where the channel error probability is very low, and occasionally switches over to a "bad" state where the channel error probability is very high. The errors seem to occur in bursts, and hence, this type of errors are called *burst errors*. Examples of channels which suffer from burst errors are the radio channel affected by signal fading, wire and cable channels affected by crosstalk and impulse noise, and magnetic recording channels affected by storage device and media malfunction. Burst errors are sometimes also called erasures.

Channels may suffer from a combination of these two types of errors, and may require special error control mechanisms. A commonly used technique to guard against burst errors is *interleaving*, which has the effect of spreading the burst of errors so that they resemble random errors. In this book, we will not concern ourselves with burst errors. Interested readers can refer to [285].

Packet loss during congested traffic conditions on high-speed networks is a special type of error which can be controlled by coding. It can be regarded as an erasure of known size and location, occurring due to limited resources (buffer overflow) in the network rather than noise on the channel.

### 1.4.2 Types of Error Control Codes

There are two basic types of codes: block and convolutional. Block codes independently encode fixed length blocks of the information. Convolutional codes differs from block codes in that each encoding operation depends on current as well as a number of past information bits. This number is known as the *constraint length* of the code. It is desirable to have large constraint lengths (or block lengths) for efficient encoding. However, the decoder complexity increases in proportion to the constraint (or block) length. Convolutional codes perform as good as block codes in most practical applications, and in some cases offer a clear performance advantage over block codes [452, 324, 46, 42].

Codes can also be categorized as error-detecting and error-correcting codes. Both block codes and convolutional codes can be used as error-detecting or error-correcting codes. Further, both block and convolutional codes can be designed to have simultaneous error-detecting and correcting capabilities. Yet another coding technique using multiple encoding and decoding stages was introduced by Forney [144], and is known as *concatenated coding* (see Appendix D).

### 1.4.3 Types of Error Control Protocols

In many communication systems, only error-detecting capability is required, and messages received with errors are retransmitted. Such schemes are quite common and are known as *automatic-repeat-request* (ARQ) schemes. On the other hand, systems involving long round-trip delays may suffer heavy throughput degradation due to retransmissions, and it may be desirable to correct errors at the receiver without asking for retransmissions. Schemes involving only error-correction are known as *forward error correction* (FEC) schemes. FEC is the only error control alternative when no feedback channel is available. A third possibility is a combination of ARQ and FEC, and is known as *hybrid* ARQ.

It is instructive to classify the error control techniques with respect to the amount of coding redundancy used by each of them. Given some measure of noise that the channel introduces, it is possible to determine the amount of coding redundancy necessary to achieve a certain performance (specified by the desirable probability of error after decoding). Stationary channels have noise characteristics that stay almost constant over time, while nonstationary channels have noise behavior that varies with time. This suggests two categories of error control protocols:

- Fixed-Redundancy Protocols: The two types of fixed-redundancy protocols are fixed-redundancy ARQ and fixed-redundancy hybrid ARQ. Conventional ARQ protocols such as stop-and-wait, go-back-N, and selective-repeat belong to the first category, while modifications of these ARQ protocols that include embedded FEC belong to the second.

- Variable-Redundancy Protocols: Variable-redundancy error control protocols are typically hybrid ARQ protocols. These protocols initially use as much redundancy as do conventional ARQ schemes. If errors are detected at the receiver and retransmissions are requested, increasingly larger amounts of coding redundancy corresponding to a message is sent to the receiver. Thus, these types of schemes are able to vary the coding redundancy in proportion to the changing channel noise characteristics and are good for nonstationary channels.

### *Error Recovery in High-Speed Networks*

There are two major classes of errors in high-speed networks, (e.g., *asynchronous transfer mode* (ATM) networks): random errors, and errors due to lost packets resulting from buffer overflow. It is expected that with fiber optic media as the basic building blocks of the ATM networks, the channel error rate for random errors will be very low. Hence, packet loss due to buffer overflows during congested traffic conditions is expected to be the primary limitation in network performance. Efficient protocols must be designed to provide reliable communication using error control techniques to recover lost packets.

Efforts to design better end-to-end communication protocols for high-speed networks have traditionally focused on using ARQ schemes. ARQ schemes require buffering of the messages at the transmitter and the receiver until the correctness of received messages can be established by using the feedback channel and a closed loop type feedback approach. The performance of end-to-end protocols degrades rapidly if the underlying error control scheme is based on error detection and retransmission of packets that make up a long message or a *protocol data unit* (PDU). This phenomenon is more pronounced in high-speed networks since the ratio of packet transmission time to propagation delay is very small. Using FEC schemes to reduce the number of retransmissions necessary for error control is a lucrative alternative [20, 50, 49, 347, 412].

## 1.5 CONVOLUTIONAL CODING

Consider the circuit shown on Figure 1.3. The square boxes are delay elements (or memory elements), and the circles are modulo-2 adders. The information

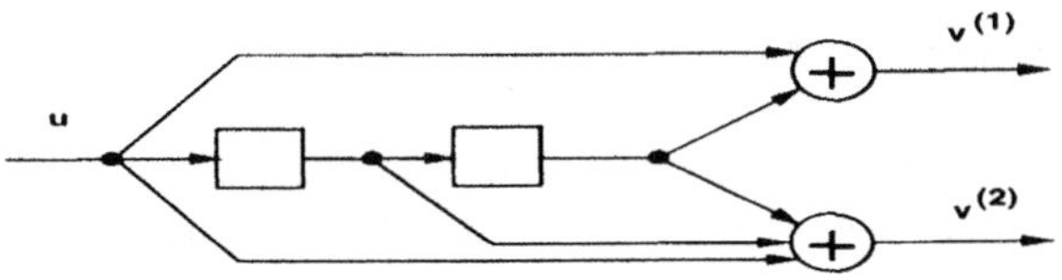


**Figure 1.3** A convolutional encoder.

sequence **u** is shifted in from the left one bit at a time, and two encoded sequences $\mathbf{v}^{(1)}$ and $\mathbf{v}^{(2)}$ are the outputs of the modulo-2 adders. Thus, at time unit $i$, the current information bit is $u_i$, the contents of the two delay elements are $u_{i-1}$ and $u_{i-2}$, respectively, and the two outputs are $v_i^{(1)}$ and $v_i^{(2)}$. The first output $v_i^{(1)}$ is the modulo-2 sum of $u_i$ and $u_{i-2}$, while the second output $v_i^{(2)}$ is the modulo-2 sum of $u_i$, $u_{i-1}$, and $u_{i-2}$. The sequence of encoded bits is multiplexed and transmitted over the channel.

Consider the information sequence $\mathbf{u} = 1101001$. Assume that the delay elements initially contain 0's. Then, at time $i = 0$, the information bit $u_0 = 1$ generates two encoded bits $v_0^{(1)} = 1$ and $v_0^{(2)} = 1$. At time $i = 1$, $u_0 = 1$ is shifted into the first delay element, and the zero in it is shifted into the second delay element. The current information bit $u_1 = 1$ generates two encoded bits $v_0^{(1)} = 1$ and $v_0^{(2)} = 0$. At time $i = 2$, $u_1 = 1$ is shifted into the first delay element, and $u_0 = 1$ in it is shifted into the second delay element. The current information bit $u_2 = 0$ generates two encoded bits $v_0^{(1)} = 1$ and $v_0^{(2)} = 0$. Continuing in this way, the entire encoded sequence is $\mathbf{v} = 11\ 10\ 10\ 00\ 01\ 11\ 11\ 01\ 11$. Note that the last two pairs of encoded bits are obtained by shifting in 0's after $u_6$ at times $i = 7$ and 8.

In general, at time $i$, a block of $k$ information bits is shifted in and a block of $n$ encoded bits is generated. As mentioned earlier, the basic difference between block and convolutional coding is that in block coding, the encoded block at time $i$ depends *only* on the information block at time $i$, while in convolutional coding, the encoded block at time $i$ depends not only on the information block at time $i$ but also on $m$ previous information blocks. Thus, a convolutional encoder requires memory. With this simple example, we are now ready to study the subject of convolutional coding in greater detail.

## 1.6 EXERCISES

1. Read the tutorial articles on digital communications by Sklar [423, 424], and on forward error correction (FEC) by Bhargava [46].

2. A 4-bit message is transmitted over a BSC with a transition probability $p = 0.01$. The received message is considered error-free if all four bits are correct. Compute the probability of error in the received message.

3. A 4-bit message is encoded using a single-error-correcting code to obtain a 10-bit codeword, which is transmitted over the BSC of Exercise 2. Compute the probability of error in the received 10-bit message after correction.

4. Compute the probability of error in a 4-bit uncoded received message if it was transmitted over a BSC with transition probability $p = 0.001$. Does this channel achieve the same order of probability of error in the uncoded message as the use of coding in Exercise 3?

5. Write a computer program to simulate a BSC with a given transition probability $p$ ($< 0.5$). Use a random number generator to get a number between 0.0 and 1.0, and use $p$ as a threshold to decide whether there is an error or not. Use a large bit stream of different sizes, say, 10, 1000, and 100000 bits, and run it through the BSC simulator. Compare the output with the input and compute the error rate. How does it compare with the BSC transition probability $p$?

# PART II

# DEFINITIONS, STRUCTURE, AND CLASSIFICATION

# 2

# CONVOLUTIONAL CODES AND ENCODERS

The study of convolutional codes has three aspects: encoding, error correction and decoding, and applications. The first aspect, covered in this part of the book, is concerned with the definitions, structure, and classification of convolutional codes and encoders. The second aspect deals with error correction and decoding of convolutionally encoded data. Finally, the third aspect explores applications of convolutional codes and decoding techniques. This chapter deals with the definitions and representations of convolutional codes and encoders.

## 2.1 INTRODUCTION

Channel coding is a means of processing source information prior to transmission over inherently noisy channels, with the objectives of detecting and correcting errors in the received message. A message to be sent over the communication channel can be viewed as a sequence of discrete source symbols and is called an *information sequence*. It is fed to a channel encoder, and the output of the encoder is called the *encoded sequence* (or the *codeword*). Channel coding is a method of mapping an information sequence to an encoded sequence. The set of encoded sequences corresponding to all possible information sequences drawn from a specific alphabet is called a *channel code*. Channel coding should, at the very least, allow the possibility of recovering (or *decoding*) the original information unambiguously from the *received sequence* in the absence of channel noise. That is, a one-to-one correspondence must exist between the sets of information and encoded sequences.

## *Redundancy and Structure*

A channel code systematically introduces *redundancy* in the information sequence which is used at the receiver to detect and correct errors. This is achieved by selecting the set of encoded sequences as a proper subset of the set of all possible sequences that can be transmitted (or received, assuming, for the moment, that these two sets are the same). If the entire set is used to represent the information, then it would be impossible to detect errors. Let the set of all encoded sequences be denoted by $C$. Then, the elements in $C$ are "valid" sequences (i.e., codewords), and those received sequences not in $C$ are "invalid" or *error sequences*.

A simple decoding scheme stores the information sequences corresponding to each possible received sequence in a table called a "codebook". The codebook is then consulted for decoding each received sequence. A rule is needed to decide which information sequence should be the decoded output for those received sequences that are not codewords. The size of the codebook is exponentially proportional to the length of the received sequences, which in turn depends on the length of the information sequences. It is quite clear that even for binary information sequences of moderate lengths, the size of the codebook may become prohibitively large and may significantly increase the complexity of decoding. Hence, it is necessary to explore ways to choose codes for which the decoding process is tractable. One possible approach is to introduce *structure* in codewords which can be exploited to simplify the decoding procedure.

## *Linear Codes*

In practice, the requirements of structure and redundancy can both be met by restricting the channel code to be a *linear code*. The set of discrete source symbols, called the *source alphabet*, is assumed to be the finite field $GF(q)$. Sequences of source symbols form a (possibly infinite) set of information sequences. Hence, the set of information sequences is a type of "meta-alphabet". It is desirable to endow this set with the structure of a field. For the moment, we will simply assume that the source alphabet is a field $F$ (finite or infinite). The set of $n$-tuples of elements of $F$, denoted by $F^n$, is an n-dimensional vector space over the field $F$. The vector space $F^n$ is a good choice for the set of all possible transmitted (or received) sequences.

**Definition 2.1** *A linear code is a $k$-dimensional subspace of an $n$-dimensional vector space over the field of source alphabet.*

Almost all codes of practical interest are linear codes. A vector space is a group under vector addition. Hence, a linear code is often called a *group code*.

## *Linear Sequential Circuits*

As mentioned in Section 1.5, convolutional coding involves memory. Hence, an encoding device is implemented using sequential logic. A *linear sequential circuit* (LSC) is a network with a finite number of input and output terminals [163]. The signals applied to the input terminals are taken from the finite field $GF(q)$ and are applied simultaneously to all input terminals at discrete instants of time. The network in the LSC consists of an interconnection of a finite number of primitive components. Three types of primitive components are considered: *adders* to implement addition modulo-$q$, *scalers* to implement multiplication modulo-$q$, and *delay elements* to delay (or store) an input signal for one time unit. The number of delay elements in an LSC is called the *dimension* of the LSC.

The basic circuit variables of an LSC are the inputs, outputs, and the states. If the LSC has more than one input, output, or state, they can be represented by input vector, output vector, and state vector, respectively. A state variable at a given time unit is the output of a delay element at that time. If the total number of delay elements in the LSC is $\nu$, it is called a $\nu$-dimensional LSC. A $\nu$-dimensional space of state vectors is called the state space of the LSC.

Two commonly used canonical forms of a linear system are the *controller canonical form* and the *observer canonical form.* In the controller canonical form of a single input single output linear system, the delay elements form a shift register; the output is a linear function of the input and the shift register contents, and the input to the shift registers is a linear function of the output and the shift register contents. The output can be expressed as the multiplication of the input and a *rational transfer function.* The input $u(D)$ and output $v(D)$ of an LSC are expressed as polynomials in an indeterminate $D$. A *delayfree* polynomial is one with a nonzero constant term. A *rational function* $g(D) = p(D)/q(D)$ is a ratio of two polynomials $p(D)$ and $q(D)$, where $q(D) \neq 0$. A rational function is called *realizable* if $q(D)$ is delayfree. In general, a matrix $\mathbf{G}(D)$ whose entries are rational functions is called a *rational transfer function matrix.* A rational transfer function matrix whose entries are realizable functions is called realizable. In practice, a rational transfer function matrix is realized by an LSC.[1]

[1] See [147, 232, 225] for a detailed description of canonical forms of an LSC. See Appendix A for definitions of matrices, polynomials, rational functions, etc.

## 2.2 DEFINING CONVOLUTIONAL CODES AND ENCODERS

Newcomers to the subject of convolutional coding often experience an initial confusion about the definition of a convolutional code, largely due to many different definitions found in the literature. Many authors have used the term convolutional code in place of convolutional encoder, and have attributed to a convolutional code properties that are, in fact, the properties of an encoder. The code and the encoder are closely related and yet distinct. In order to avoid the confusion, it is necessary to understand the distinction between the following related terms: a convolutional encoder, a convolutional encoding operation, and a convolutional code. The *encoder* is a device or a machine. The *encoding operation* is facilitated by the encoder. The objects that are encoded are *information sequences*, and the corresponding outputs are the *encoded sequences.* Finally, a *convolutional code* is the set of encoded sequences corresponding to all possible information sequences. In conclusion, the definition of a convolutional encoder should specify the structure of the encoding machine, and that of a convolutional code should specify the structure of the set of encoded sequences.

The definitions of a convolutional encoder found in the literature also differ based on the level of generalization. Broadly speaking, there are two levels: abstract, and physically realizable. Finally, there is the question of the order in which the code and the encoder are defined. Again, two approaches are found in the literature. One of them defines the code first, as a $k$-dimensional subspace of an $n$-dimensional vector space over a suitable field, and then defines the encoder as a $k \times n$ matrix whose rows are a basis for the code. This approach is taken by Massey [317], and McEliece and Onyszchuk [326]. The other approach defines the encoder as a $k$-input $n$-output LSC, and then defines the code as the set of output sequences generated by the encoder for all possible input sequences. This approach was proposed by Forney in his landmark paper [147], and later followed by Piret [362], and Johannesson and Wan [225]. According to the first approach, the code is the row space of the encoder. In the second approach, the LSC realizes a mapping between a $k$-dimensional vector space and an $n$-dimensional vector space, whose image set is the code. When considered in juxtaposition, both these approaches are seen to describe the same objects, albeit in a different order.

## *Information and Encoded Sequences*

The level of generalization in defining convolutional codes and encoders is governed by the structure of the sets of information and encoded sequences. Elements of these sets may be drawn from the following infinite fields: the field $F((D))$ of formal Laurent series in $D$ over $F$, or the field $F(D)$ of rational functions in $D$ over $F$, where $F = GF(q)$. They may also be drawn from the ring $F[[D]]$ of formal power series in $D$ over $F$, or from the ring $F[D]$ of polynomials in $D$ over $F$.[2] In practice, the field $F$ is usually the binary field $GF(2)$. The set $F^n$ of $n$-tuples of elements of $F$ is itself a field. The field of formal Laurent series of elements of $F^n$ is denoted by $F^n((D))$. On the other hand, the set $\{F((D))\}^n$ of all $n$-tuples of elements of $F((D))$ is also a field. Here, we have adopted the convention that $\{F((D))\}^n = F^n((D))$, i.e., the $n$-tuple of sequences is the same as the sequence of $n$-tuples.

Let the information $k$-tuple at time $i$ be $\mathbf{u}_i = (u_i^{(1)}, u_i^{(2)}, \ldots, u_i^{(k)})$. The sequence of information $k$-tuples can be represented as $\mathbf{U}(D) = \sum_{i=r}^{\infty} \mathbf{u}_i D^i$ starting at some arbitrary time $r$. The input is zero for $i < r$. The sequence $\mathbf{U}(D)$ is an element of $F^k((D))$. Similarly, let the encoded $n$-tuple at time $i$ be $\mathbf{v}_i = (v_i^{(1)}, v_i^{(2)}, \ldots, v_i^{(n)})$. The sequence of encoded $n$-tuples can be represented as $\mathbf{V}(D) = \sum_{i=r}^{\infty} \mathbf{v}_i D^i$, and is an element of $F^n((D))$. On the other hand, if the information and encoded streams are considered separately, the $k$ information sequences are $\mathbf{u}^{(i)}$, $i = 1, \ldots, k$, and can be represented as the vector $\mathbf{U} = (\mathbf{u}^{(1)}, \mathbf{u}^{(2)}, \ldots, \mathbf{u}^{(k)})$. Similarly, the $n$ encoded sequences are $\mathbf{v}^{(j)}$, $j = 1, \ldots, n$, and can be represented as the vector $\mathbf{V} = (\mathbf{v}^{(1)}, \mathbf{v}^{(2)}, \ldots, \mathbf{v}^{(n)})$. Finally, $\mathbf{U}(D) = (\mathbf{u}^{(1)}(D), \ldots, \mathbf{u}^{(k)}(D))$, and $\mathbf{V}(D) = (\mathbf{v}^{(1)}(D), \ldots, \mathbf{v}^{(n)}(D))$.

## *Convolutional Code*

A *convolutional code* can now be defined as follows:

**Definition 2.2** *An $(n, k)$ convolutional code $C$ over a finite field $F$ is a $k$-dimensional subspace of $n$-dimensional vector space $\{F((D))\}^n$.*

The *code rate* of an $(n, k)$ convolutional code is $R = k/n$. A rational subcode of an $(n, k)$ convolutional code is obtained if the $n$-dimensional vector space is $\{F(D)\}^n$. Massey [317], and McEliece and Onyszchuk [326] have defined such a subcode.

[2] See Appendix A for definitions of these structures.

A convolutional code over a field $F$ can be further generalized as a linear code over a vector space, a ring, or a module, or at a more primitive level, as a group code over the additive group of $F$. This viewpoint has recently been taken by Forney and Trott [158]. They have defined a *group code* $C$ over a group $G$ as a subgroup of the group sequence space $G^{\mathcal{Z}}$, where $G^{\mathcal{Z}}$ is the set of all sequences[3] of elements of $G$ indexed by the set of integers $\mathcal{Z}$. If $G = F^n$, then we get the $(n, k)$ convolutional code over a field $F$ defined earlier.

## *Convolutional Encoder*

Now, let us consider the definition of a convolutional encoder. An encoder is a machine which accepts a $k$-tuple $\mathbf{u}_i$ as input at time $i$ and produces an $n$-tuple $\mathbf{v}_i$ as output, also at time $i$. This machine has memory, i.e., the $n$-tuple $\mathbf{v}_i$ is a function not only of the information $k$-tuple $\mathbf{u}_i$ but also of some information $k$-tuples $\mathbf{u}_j$ for $j < i$.

**Definition 2.3** *An encoder is a $k \times n$ convolutional encoder over $F$ if the mapping $F^k((D)) \to F^n((D))$ realized by the encoder can be represented by $\mathbf{V}(D) = \mathbf{U}(D)\mathbf{G}(D)$, where $\mathbf{G}(D)$ is a $k \times n$ matrix of rank $k$ with entries in the subset $F[D]$ of $F((D))$.*

It is customary to call $\mathbf{G}(D)$ as the encoder. The encoder in Definition 2.3 has *encoding rate* $R = k/n$. The terms code rate and encoding rate are used interchangeably. Definition 2.3 was given by Piret [362], who defined a convolutional code as the $F((D))$-row space of the encoder $\mathbf{G}(D)$. Piret's definitions are essentially the same as those of Forney [147].

Johannesson and Wan [225] have introduced a further distinction by defining the following separately: a convolutional transducer with its transfer function matrix, a convolutional code with its generator matrix, and a convolutional encoder with its encoding matrix. Their goal is to make a clear distinction between objects that are abstract constructs and those that are physically realizable using linear sequential circuits. In their terms, a transducer is a linear mapping specified by its transfer function matrix. Further, a convolutional code is the image set of a transducer, and its generator matrix is the transfer function matrix of the transducer. An encoding matrix is a generator matrix that is realizable and delayfree. Finally, a convolutional encoder of a convolutional code with encoding matrix $\mathbf{G}(D)$ is a realization, using an LSC, of a

[3] See Appendix A for the definition of a sequence in this context.

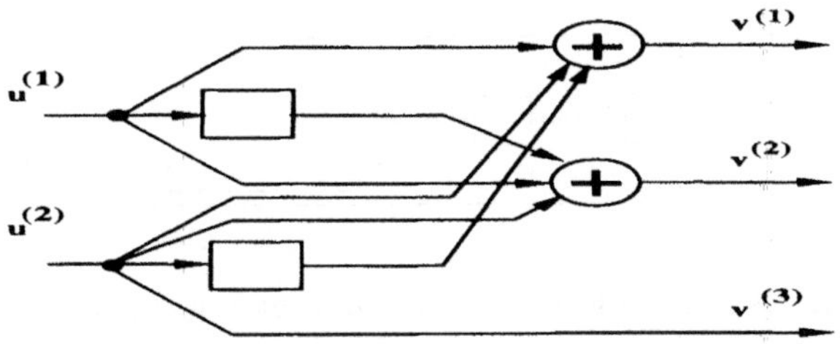


**Figure 2.1** A rate two-thirds convolutional encoder.

convolutional transducer with transfer function matrix $\mathbf{G}(D)$. They have chosen $F$ to be the binary field $GF(2)$. In this book, we do not maintain a strict distinction between a generator matrix and an encoding matrix, and refer to all of them as the encoder. Exceptions to this are explicitly mentioned.

**Example 2.1** A rate one-half convolutional encoder is shown in Figure 1.3. It consists of a shift register with two delay elements (or memory elements) shown as the square boxes, and two modulo-2 adders (or Exclusive-OR gates) shown as circles.

**Example 2.2** Figure 2.1 shows the encoder for a (3, 2) convolutional code. The encoder has two shift registers, each with an information sequence $\mathbf{u}^{(i)}$ $(i = 1,\ 2)$ as input. Similarly, there are three outputs $\mathbf{v}^{(j)}$ $(j = 1,\ 2,\ 3)$.

**Example 2.3** Figure 2.2 shows a systematic encoder for a (2, 1) convolutional code. Note that the first encoded sequence $\mathbf{v}^{(1)}$ is the same as the information sequence $\mathbf{u}$. An encoder in which the first $k$ encoded sequences are same as the $k$ information sequence is called a *systematic* encoder. Properties of systematic and nonsystematic encoders are discussed in detail in Chapter 4.

A convolutional code may be generated by many encoders. Hence, a notion of equivalence of encoders can be introduced.

**Definition 2.4** *Two convolutional encoders are equivalent if they generate the same code.*

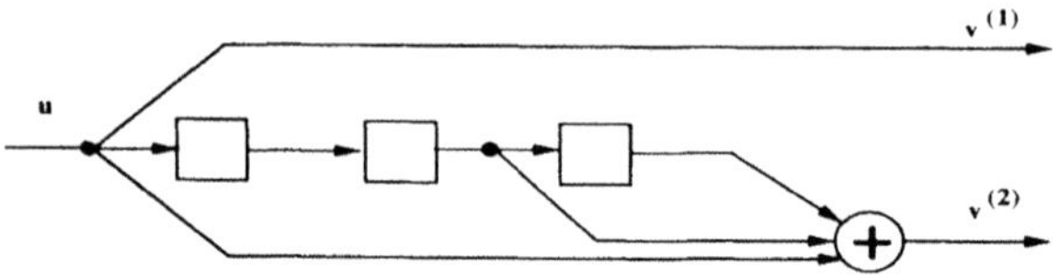


**Figure 2.2** A rate one-half systematic convolutional encoder.

The obvious question is: which among all the equivalent encoders should be chosen to generate a code? This question is addressed in the present and the next two chapters.

## *Dual Convolutional Code*

A subspace $V$ of a vector space $W$ has a dual space $V^{\perp}$ associated with it. A vector in $V$ is orthogonal to every vector in $V^{\perp}$. Further, if the $dim(V)$ is the dimension of $V$ as a vector space, then $dim(V) + dim(V^{\perp}) = dim(W)$. Hence, associated with a convolutional code $C$ is its dual $C^{\perp}$.

**Definition 2.5** *Let $C$ be an $(n,\ k)$ convolutional code. Then, its dual code $C^{\perp}$ is an $(n-k)$-dimensional subspace of $\{F((D))\}^n$ consisting of all sequences $\mathbf{V}^{\perp}(D)$ orthogonal to all encoded sequences $\mathbf{V}(D) \in C$.*

The dual code $C^{\perp}$ is itself an $(n,\ n-k)$ convolutional code. It is generated by any rate-(n-k)/n encoder $\mathbf{H}(D)$ such that $\mathbf{G}(D)\mathbf{H}^T(D) = \mathbf{0}$. The matrix $\mathbf{H}(D)$ is called the *parity-check matrix* of the code $C$. An algorithm to determine $\mathbf{H}(D)$, given $\mathbf{G}(D)$, is discussed later in this chapter.

## *Convolutional Syndrome-Former*

Forney [149] has termed the transpose of the parity-check matrix $\mathbf{H}(D)$ as the *syndrome-former*:

**Definition 2.6** *The $n$-input, $(n-k)$-output linear sequential circuit whose transfer function matrix is $\mathbf{H}^T(D)$ is called a syndrome-former, and has the property that $\mathbf{V}(D)\mathbf{H}^T(D) = \mathbf{0}$ if and only if $\mathbf{V}(D) \in C$.*

That is, when the $n$ outputs of an encoder $\mathbf{G}(D)$ are connected to the $n$ inputs of the corresponding syndrome-former $\mathbf{H}^T(D)$, the $n-k$ outputs of the syndrome-former are zero for all time.

Let $\mathbf{R}(D)$ be the received sequence when the codeword $\mathbf{V}(D)$ is transmitted over a channel. The sequence $\mathbf{R}(D) \in \{F((D))\}^n$ may be different from the transmitted codeword $\mathbf{V}(D)$ since the channel may introduce errors. Let $\mathbf{E}(D)$ be the error sequence. Then,

$$\mathbf{R}(D) = \mathbf{V}(D) + \mathbf{E}(D). \tag{2.1}$$

The syndrome-former $\mathbf{H}^T(D)$ is useful in forming the *syndrome*[4] sequence corresponding to a received sequence $\mathbf{R}(D)$.

**Definition 2.7** *The syndrome vector* $\mathbf{S}(D)$ *is defined as*

$$\mathbf{S}(D) = \mathbf{R}(D)\mathbf{H}^T(D). \tag{2.2}$$

Substituting from (2.1) in (2.2), we get

$$\mathbf{S}(D) = \mathbf{V}(D)\mathbf{H}^T(D) + \mathbf{E}(D)\mathbf{H}^T(D). \tag{2.3}$$

But, $\mathbf{V}(D)\mathbf{H}^T(D) = \mathbf{0}$, and hence

$$\mathbf{S}(D) = \mathbf{E}(D)\mathbf{H}^T(D). \tag{2.4}$$

Thus, the syndrome depends only on the errors introduced by the channel and not on the transmitted codeword. Examples of syndrome-formers are included later in this chapter.

## 2.3 REPRESENTATIONS OF CONVOLUTIONAL ENCODERS

Shift register based implementations of convolutional encoders were considered in the previous section. The convolutional encoder was also represented in a mathematical form. In this section, many such representations are given.

[4]The term "syndrome" in medical terminology means a number of symptoms of a disease. In the context of coding, the disease is the errors and the symptom is the syndrome. This coinage is due to Hagelbarger [324].

The convolutional encoding operation can be specified precisely using various representations of the encoder. The encoding operation can also be depicted graphically in a variety of ways. Throughout the rest of this book, an "encoder" will mean a convolutional encoder, unless otherwise specified.

### 2.3.1 Analytical Representations

Many different yet related analytical representations of convolutional encoders and of the convolutional encoding operation are described.

#### *Discrete Generator Sequences*

A rate-$k/n$ convolutional encoder is represented by $nk$ generator sequences $\mathbf{g}_i^{(j)} = (g_{i,0}^{(j)}, g_{i,1}^{(j)}, \ldots, g_{i,m}^{(j)})$ $i = 1, \ldots, k$, $j = 1, \ldots, n$. The convolutional encoding operation, as the name implies, is the discrete convolution of the information sequence with the generator sequences, and is expressed as

$$\mathbf{v}^{(j)} = \sum_{i=1}^{k} \mathbf{u}^{(i)} * \mathbf{g}_i^{(j)}, \quad j = 1, 2, .., n. \tag{2.5}$$

Thus, each of the $n$ encoded sequences $\mathbf{v}^{(j)} = (v_0^{(j)}, v_1^{(j)}, v_2^{(j)}, \ldots), j = 1, \ldots, n$ may depend upon each of the $k$ information sequences $\mathbf{u}^{(i)} = (u_0^{(i)}, u_1^{(i)}, u_2^{(i)}, \ldots)$, $i = 1, \ldots, k$. The generator sequences define the nature of this dependence. The composite generator sequence for the $i$th input is

$$\mathbf{g}_i = (g_{i,0}^{(1)}, g_{i,0}^{(2)}, \ldots, g_{i,0}^{(n)}, g_{i,1}^{(1)}, \ldots, g_{i,1}^{(n)}, \ldots, g_{i,m}^{(1)}, \ldots, g_{i,m}^{(n)}). \tag{2.6}$$

For $k = 1$, the composite generator sequence is simply

$$\mathbf{g} = (g_0^{(1)}, g_0^{(2)}, \ldots, g_0^{(n)}, g_1^{(1)}, \ldots, g_1^{(n)}, \ldots, g_m^{(1)}, \ldots, g_m^{(n)}). \tag{2.7}$$

An encoder with parameters $n$, $k$, and $m$ is called an $(n, k, m)$ encoder.

**Example 2.4** The (2, 1, 2) encoder in Figure 1.3 has generator sequences $\mathbf{g}^{(1)} = 101$ and $\mathbf{g}^{(2)} = 111$, and composite generator sequence $\mathbf{g} = 11\ 01\ 11$. Note that the generator sequences specify the inputs to each modulo-2 adder.

For the encoder of Figure 1.3, the encoding operation can be written as the following difference equations:

$$\begin{aligned} v_t^{(1)} &= u_t \qquad\qquad + u_{t-2} \\ v_t^{(2)} &= u_t + u_{t-1} + u_{t-2}, \end{aligned}$$

where we define $u_t = 0$ for $t < 0$.

## *Discrete Semi-Infinite Generator Matrix*

The discrete convolution in (2.5) can be more compactly expressed as a matrix multiplication. The $k$ information sequences can be interleaved to obtain the composite information sequence $\mathbf{u} = (\mathbf{u}_0, \mathbf{u}_1, \ldots) = (u_0^{(1)}, u_0^{(2)}, \ldots, u_0^{(k)}, u_1^{(1)}, \ldots, u_1^{(k)}, \ldots)$, where $\mathbf{u}_t = (u_t^{(1)}, \ldots, u_t^{(k)})$ is the information block at time $t$. Similarly, the composite encoded sequence (also called a codeword) obtained by interleaving the $n$ encoded sequences is $\mathbf{v} = (\mathbf{v}_0, \mathbf{v}_1, \ldots) = (v_0^{(1)}, v_0^{(2)}, \ldots, v_0^{(n)}, v_1^{(1)}, \ldots, v_1^{(n)}, \ldots)$, where $\mathbf{v}_t = (v_t^{(1)}, \ldots, v_t^{(n)})$ is the encoded block at time $t$. Now, the convolutional encoding is given by

$$\mathbf{v} = \mathbf{u}\mathbf{G}. \tag{2.8}$$

Here, the matrix $\mathbf{G}$ is a semi-infinite generator matrix

$$\mathbf{G} = \begin{bmatrix} \mathbf{G}_0 & \mathbf{G}_1 & \mathbf{G}_2 & \cdots & \mathbf{G}_m & & & \\ & \mathbf{G}_0 & \mathbf{G}_1 & \cdots & \mathbf{G}_{m-1} & \mathbf{G}_m & & \\ & & \mathbf{G}_0 & \cdots & \mathbf{G}_{m-2} & \mathbf{G}_{m-1} & \mathbf{G}_m & \\ & & & \ddots & & & & \ddots \end{bmatrix}, \tag{2.9}$$

where blanks indicate zeros, and

$$\mathbf{G}_l = \begin{bmatrix} g_{1,l}^{(1)} & g_{1,l}^{(2)} & \cdots & g_{1,l}^{(n)} \\ g_{2,l}^{(1)} & g_{2,l}^{(2)} & \cdots & g_{2,l}^{(n)} \\ \vdots & \vdots & & \vdots \\ g_{k,l}^{(1)} & g_{k,l}^{(2)} & \cdots & g_{k,l}^{(n)} \end{bmatrix}. \tag{2.10}$$

The generator matrix $\mathbf{G}$ is a semi-infinite matrix since the information sequence may be infinite. In general, an encoded block $\mathbf{v}_t$ is given by

$$\mathbf{v}_t = \mathbf{u}_t\mathbf{G}_0 + \mathbf{u}_{t-1}\mathbf{G}_1 + \cdots + \mathbf{u}_{t-m}\mathbf{G}_m, \tag{2.11}$$

where $\mathbf{u}_{t-i} = \mathbf{0}$ for $t < i$.

For a (2, 1, m) encoder, we have $\mathbf{G}_l = \begin{bmatrix} g_l^{(1)} & g_l^{(2)} \end{bmatrix}$, and the generator matrix is

$$\mathbf{G} = \begin{bmatrix} g_0^{(1)}g_0^{(2)} & g_1^{(1)}g_1^{(2)} & g_2^{(1)}g_2^{(2)} & \cdots & g_m^{(1)}g_m^{(2)} & & \\ & g_0^{(1)}g_0^{(2)} & g_1^{(1)}g_1^{(2)} & \cdots & g_{m-1}^{(1)}g_{m-1}^{(2)} & g_m^{(1)}g_m^{(2)} & \\ & & g_0^{(1)}g_0^{(2)} & \cdots & g_{m-2}^{(1)}g_{m-2}^{(2)} & g_{m-1}^{(1)}g_{m-1}^{(2)} & g_m^{(1)}g_m^{(2)} \\ & & & \ddots & & & \ddots \end{bmatrix}. \tag{2.12}$$

**Example 2.5** The semi-infinite generator matrix **G** for the (2, 1, 2) encoder in Figure 1.3 is

$$\mathbf{G} = \begin{bmatrix} 11 & 01 & 11 & & & \\ & 11 & 01 & 11 & & \\ & & 11 & 01 & 11 & \\ & & & 11 & 01 & 11 \\ & & & & \ddots & & \ddots \end{bmatrix}.$$

The encoded sequence corresponding to the information sequence $\mathbf{u} = 1101001$ is obtained by performing the following matrix multiplication:

$$(1101001)\begin{bmatrix} 11 & 01 & 11 & & & & & & \\ & 11 & 01 & 11 & & & & & \\ & & 11 & 01 & 11 & & & & \\ & & & 11 & 01 & 11 & & & \\ & & & & 11 & 01 & 11 & & \\ & & & & & 11 & 01 & 11 & \\ & & & & & & 11 & 01 & 11 \end{bmatrix},$$

which yields $\mathbf{v} = 11\ 10\ 10\ 00\ 01\ 11\ 11\ 01\ 11$. Note that this sequence is the same as that obtained in Section 1.5.

In Example 2.5, an information sequence of length $N = 7$ bits was encoded into a codeword of length $M = 18$ bits. The last four bits in the codeword correspond to the $m = 2$ zeros used for termination. In general, by using an $(n, k, m)$ encoder, an information sequence of length $N$ blocks, i.e., $k \cdot N$ bits is encoded into a codeword of length $M = (N + m)$ blocks, i.e., $n(N + m)$ bits, where the last $n \cdot m$ bit are generated by the $m$ all-zero information blocks appended to the information sequence for termination.

The first row of **G** is the composite generator sequence $\mathbf{g} = 11\ 01\ 11$. The second row is obtained by shifting the first row two positions to the right. Each successive row is the previous row shifted two positions to the right. In general, for an $(n,\ k,\ m)$ encoder, successive sets of $k$ rows are shifted to the right by $n$ positions. This captures the shifting of information sequence in the convolution operation implemented using a shift register.

**Example 2.6** The semi-infinite generator matrix **G** for the (3, 2, 1) encoder in Figure 2.1 is

$$\mathbf{G} = \begin{bmatrix} 110 & 010 & & \\ 111 & 100 & & \\ & 110 & 010 & \\ & 111 & 100 & \\ & & \ddots & \ddots \end{bmatrix}.$$

## *Discrete Semi-Infinite Parity-Check Matrix*

Given **G**, the discrete semi-infinite *parity-check matrix* **H** is defined as a matrix with the property that $\mathbf{G}\mathbf{H}^T = \mathbf{0}$. For a $(2,\ 1,\ m)$ encoder with generator sequences $\mathbf{g}^{(1)}$ and $\mathbf{g}^{(2)}$, the $(m+1)th$ row of **H** is called the *parity-check sequence* **h**, and is given by

$$\mathbf{h} = (g_m^{(2)},\ g_m^{(1)},\ g_{m-1}^{(2)},\ g_{m-1}^{(1)},\ \ldots,\ g_0^{(2)},\ g_0^{(1)}). \tag{2.13}$$

Notice that **h** is the reverse of **g** for $(2,\ 1,\ m)$ encoders. The matrix $\mathbf{H}^T$ is called the *syndrome-former*, and is given by

$$\mathbf{H}^T = \begin{bmatrix} g_0^{(2)} & g_1^{(2)} & \cdots & g_m^{(2)} & & \\ g_0^{(1)} & g_1^{(1)} & \cdots & g_m^{(1)} & & \\ & g_0^{(2)} & \cdots & g_{m-1}^{(2)} & g_m^{(2)} & \\ & g_0^{(1)} & \cdots & g_{m-1}^{(1)} & g_m^{(1)} & \\ & & & \vdots & \vdots & \ddots \\ & & & g_0^{(2)} & g_1^{(2)} & \\ & & & g_0^{(1)} & g_1^{(1)} & \\ & & & & g_0^{(2)} & \\ & & & & g_0^{(1)} & \\ & & & & & \ddots \end{bmatrix}. \tag{2.14}$$

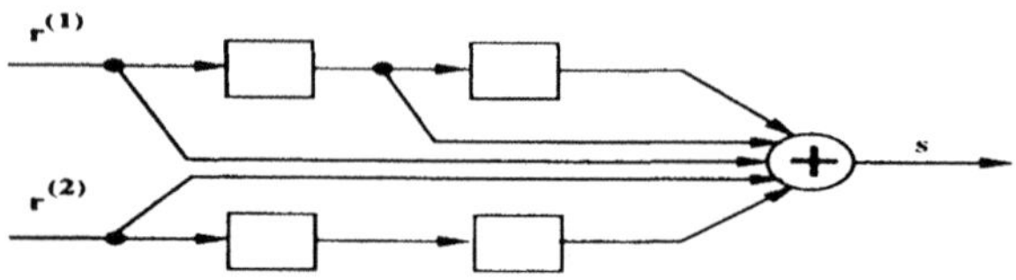


**Figure 2.3** A rate one-half syndrome-former.

**Example 2.7** The semi-infinite syndrome former matrix $\mathbf{H}^T$ for the (2, 1, 2) encoder with generator sequences $\mathbf{g}^{(1)} = 101$ and $\mathbf{g}^{(2)} = 111$ is

$$\mathbf{H}^T = \begin{bmatrix} 1 & 1 & 1 & & & \\ 1 & 0 & 1 & & & \\ & 1 & 1 & 1 & & \\ & 1 & 0 & 1 & & \\ & & 1 & 1 & 1 & \\ & & 1 & 0 & 1 & \\ & & & 1 & 1 & \ddots \\ & & & 1 & 0 & \\ & & & & 1 & \\ & & & & 1 & \\ & & & & & \ddots \end{bmatrix}.$$

The parity-check sequence for this encoder is $\mathbf{h} = 11\ 10\ 11$.

The syndrome sequence can now be generated using the syndrome-former $\mathbf{H}^T$ as follows:

$$\mathbf{s} = \mathbf{r}\mathbf{H}^T.$$

This can set up as a matrix multiplication. Given the length of the received sequence, the syndrome former can be terminated such that it has the same number of rows as the length of the received sequence. Then, a finite length syndrome sequence is obtained.

We can think of the syndrome-former as a shift register circuit. A rate one-half encoder has one input and two outputs, and it implements the matrix equation $\mathbf{v} = \mathbf{u}\mathbf{G}$. A rate one-half syndrome former has two inputs $\mathbf{r}^{(1)}$ and $\mathbf{r}^{(2)}$ and one output $\mathbf{s}$. It implements the matrix equation $\mathbf{s} = \mathbf{r}\mathbf{H}^T$. Figure 2.3 shows the shift-register implementation of the syndrome-former for the rate one-half encoder of Example 2.4.

## *Generator Polynomials*

The generator sequences $\mathbf{g}_i^{(j)}$, $i = 1, \ldots, k$, $j = 1, \ldots, n$ are finite and can be represented as polynomials of finite degree in the delay operator $D$ (also called $D$-transforms). The generator polynomials for an $(n,\ k,\ m)$ encoder are $\mathbf{g}_i^{(j)}(D) = g_{i,0}^{(j)} + g_{i,1}^{(j)} D + \ldots + g_{i,m}^{(j)} D^m$ for $i = 1, \ldots, k$, $j = 1, \ldots, n$. The information and encoded sequences can be expressed using $D$-transforms as $\mathbf{u}^{(i)}(D) = u_0^{(i)} + u_1^{(i)} D + u_2^{(i)} D^2 + \cdots$ and $\mathbf{v}^{(j)}(D) = v_0^{(j)} + v_1^{(j)} D + v_2^{(j)} D^2 + \cdots$, respectively. Since convolution in time-domain is equivalent to multiplication in the transform-domain, the encoding operation now becomes

$$\mathbf{v}^{(j)}(D) = \sum_{i=1}^{k} \mathbf{u}^{(i)}(D)\mathbf{g}_i^{(j)}(D). \tag{2.15}$$

The polynomial multiplication is carried out over the ring of polynomials $F[D]$ if the information and encoded sequences are polynomials as well. If these sequences are expressed as Laurent series in $D$ over $F$, then the multiplication also has to be appropriately generalized.

**Example 2.8** The generator polynomials of the (2, 1, 2) encoder of Figure 1.3 are $\mathbf{g}^{(1)}(D) = 1 + D^2$, and $\mathbf{g}^{(2)}(D) = 1 + D + D^2$. Similarly, the generator polynomials of the (3, 2, 1) encoder of Figure 2.1 are $\mathbf{g}_1^{(1)}(D) = 1$, $\mathbf{g}_1^{(2)}(D) = 1 + D$, $\mathbf{g}_1^{(3)}(D) = 0$, $\mathbf{g}_2^{(1)}(D) = 1 + D$, $\mathbf{g}_2^{(2)}(D) = 1$, and $\mathbf{g}_2^{(3)}(D) = 1$. Finally, the generator polynomials of the (2, 1, 2) systematic encoder of Figure 2.2 are $\mathbf{g}^{(1)}(D) = 1$, and $\mathbf{g}^{(2)}(D) = 1 + D^2 + D^3$.

## *Polynomial Generator Matrix*

In the transform-domain, an $(n,\ k,\ m)$ encoder can be represented by a $k \times n$ matrix $\mathbf{G}(D)$, called the *polynomial generator matrix* and given by

$$\mathbf{G}(D) = \begin{bmatrix} \mathbf{g}_1^{(1)}(D) & \mathbf{g}_1^{(2)}(D) & \cdots & \mathbf{g}_1^{(n)}(D) \\ \mathbf{g}_2^{(1)}(D) & \mathbf{g}_2^{(2)}(D) & \cdots & \mathbf{g}_2^{(n)}(D) \\ \vdots & \vdots & & \vdots \\ \mathbf{g}_k^{(1)}(D) & \mathbf{g}_k^{(2)}(D) & \cdots & \mathbf{g}_k^{(n)}(D) \end{bmatrix}. \tag{2.16}$$

Here, each entry $\mathbf{g}_i^{(j)}(D)$ is a polynomial. The encoding operation is

$$\mathbf{V}(D) = \mathbf{U}(D)\mathbf{G}(D). \tag{2.17}$$

The vector $\mathbf{U}(D)$ has $k$ components, namely, the information polynomials $\mathbf{u}^{(i)}(D)$, $i = 1, \ldots, k$. Similarly, the vector $\mathbf{V}(D)$ has $n$ components $\mathbf{v}^{(j)}(D)$, $j = 1, \ldots, n$.

The generator polynomials $\mathbf{g}_i^{(j)}(D)$ for any given $i$ and for all $j = 1, \ldots, n$ capture the influence of the $i$th input on all $n$ outputs. The generator polynomial (or polynomials) of highest degree for a given $i$ determines the maximum number of past information symbols of the $i$th input stream that need to be stored in the $i$th shift register. The measures of the "length" of an $(n, k, m)$ encoder are captured in the following definitions.

**Definition 2.8** *The constraint length for the ith input is*

$$\nu_i = \max_{1 \leq j \leq n} [deg\,(\mathbf{g}_i^{(j)}(D))]. \tag{2.18}$$

**Definition 2.9** *The overall constraint length is*

$$\nu = \sum_{i=1}^{k} \nu_i. \tag{2.19}$$

**Definition 2.10** *The memory order is*

$$m = \max_{1 \leq i \leq k} \nu_i. \tag{2.20}$$

**Definition 2.11** *The output constraint length is*

$$n_A = n(m+1). \tag{2.21}$$

**Definition 2.12** *The input constraint length is*

$$K = \nu + k. \tag{2.22}$$

The overall constraint length is the total number of memory or delay elements required in a shift register implementation of the encoder. The memory order is the maximum degree of an input generator polynomial. For an $(n, 1, m)$ encoder, we have $k = 1$ and $\nu = \nu_1 = m$. The output constraint length is equal to the number of encoded bits affected by a single information bit. The input constraint length is equal to the number of information bits that affect any output.

The constraint length of a convolutional encoder is an important parameter. For each of the definitions above, symbols that are most commonly used in the literature have been chosen. However, it must be noted that the constraint length, the output constraint length, the input constraint length, and the memory order have each been defined as just the "constraint length" by different authors.

**Example 2.9** The polynomial generator matrix for the (2, 1, 2) encoder of Figure 1.3 is $\mathbf{G}(D) = \begin{bmatrix} \mathbf{g}^{(1)}(D) & \mathbf{g}^{(2)}(D) \end{bmatrix} = \begin{bmatrix} 1+D^2 & 1+D+D^2 \end{bmatrix}$. This encoder has $\nu = 2$. Each information bit affects three sets of output bits. Hence, the output constraint length of the encoder is $n_A = 6$. Its input constraint length is $K = \nu + k = 3$, and its memory order is $m = \nu = 2$. The polynomial generator matrix for the (3, 2, 1) encoder of Figure 2.1 is

$$\mathbf{G}(D) = \begin{bmatrix} \mathbf{g}_1^{(1)}(D) & \mathbf{g}_1^{(2)}(D) & \mathbf{g}_1^{(3)}(D) \\ \mathbf{g}_2^{(1)}(D) & \mathbf{g}_2^{(2)}(D) & \mathbf{g}_2^{(3)}(D) \end{bmatrix} = \begin{bmatrix} 1 & 1+D & 0 \\ 1+D & 1 & 1 \end{bmatrix}.$$

This encoder has $\nu_1 = 1$, $\nu_2 = 1$, and $\nu = \nu_1 + \nu_2 = 2$. The output constraint length of the encoder is $n_A = 6$, the input constraint length is $K = \nu + k = 4$, and the memory order is $m = 1$.

The state of an encoder is the contents of the shift registers. This is called the *physical* state, and is dependent on the particular realization of the encoder. If $\mathbf{G}(D)$ is polynomial, then the dimension of the physical state space of its controller canonical form is equal to the overall constraint length $\nu$.

The encoder states can be also defined to be independent of the realization [147, 149]. Let $P$ be a projection operator that truncates sequences to end at time zero, and $Q = 1 - P$ the projection operator that truncates sequences to start at time 1. That is, for a sequence $\mathbf{x}(D)$,

$$\mathbf{x}(D)P = \begin{cases} x_d D^d + \ldots + x_0 & d \leq 0 \\ 0 & d > 0 \end{cases} \tag{2.23}$$

$$\mathbf{x}(D)Q = x_1 D + x_2 D^2 + \ldots \tag{2.24}$$

Let $\mathbf{U}(D)$ be any $k$-tuple of inputs. The *abstract state* (at time 1) associated with $\mathbf{U}(D)$ is $\zeta(\mathbf{U}(D)) = \mathbf{U}(D)P\mathbf{G}(D)Q$, i.e., the $n$-tuple of outputs at time 1 and later that are due to that part of $\mathbf{U}(D)$ that occurs upto time 0. The abstract state depends only on $\mathbf{G}(D)$ and not on the realization. The set of all abstract states is the abstract state space denoted by $\Sigma_{\mathbf{G}}$. It is a linear vector space over $F$ since $P$, $Q$, and $\mathbf{G}(D)$ are all linear over $F$. Distinct abstract states spring from distinct physical states.

## *Polynomial Parity-Check Matrix*

An $(n, k)$ convolutional code $C$ has an $(n, n-k)$ dual convolutional code $C^{\perp}$ associated with it. The dual code is generated by any rate-(n-k)/n encoder $\mathbf{H}(D)$ such that $\mathbf{G}(D)\mathbf{H}^T(D) = \mathbf{0}$. If the entries of $\mathbf{H}(D)$ are all polynomials, then $\mathbf{H}(D)$ is called the *polynomial* parity-check matrix of the code $C$, and the matrix $\mathbf{H}^T(D)$ is called the *polynomial* syndrome-former.

**Example 2.10** The polynomial parity-check matrix for the (2, 1, 2) encoder of Figure 1.3 is $\mathbf{H}(D) = \left[\begin{array}{cc} \mathbf{g}^{(2)}(D) & \mathbf{g}^{(1)}(D) \end{array}\right] = \left[\begin{array}{cc} 1+D+D^2 & 1+D^2 \end{array}\right]$, and the syndrome former matrix is

$$\mathbf{H}^T(D) = \left[\begin{array}{c} \mathbf{g}^{(2)}(D) \\ \mathbf{g}^{(1)}(D) \end{array}\right] = \left[\begin{array}{c} 1+D+D^2 \\ 1+D^2 \end{array}\right].$$

It is easy to verify that $\mathbf{G}(D)\mathbf{H}^T(D) = 0$.

Given an encoder $\mathbf{G}(D)$, it is possible to find its parity-check matrix $\mathbf{H}(D)$, as outlined in the algorithm below. The algorithm is based on application of the invariant factor theorem for matrices over a principal ideal domain, which is included in Appendix A. The invariant factor decomposition of $\mathbf{G}(D)$ is

$$\mathbf{G}(D) = \mathbf{A}(D)\Lambda(D)\mathbf{B}(D), \tag{2.25}$$

where $\mathbf{A}(D)$ and $\mathbf{B}(D)$ are unimodular matrices, i.e., they are square polynomial matrices with determinant equal to 1 and with polynomial inverses. The inverse of $\mathbf{G}(D)$ is

$$\mathbf{G}^{-1}(D) = \mathbf{B}^{-1}(D)\Lambda^{-1}(D)\mathbf{A}^{-1}(D). \tag{2.26}$$

Let $\mathbf{B}_1(D)$ be the $n \times (n-k)$ matrix composed of the last $(n-k)$ columns of $\mathbf{B}^{-1}(D)$. Then, a parity-check matrix $\mathbf{H}(D)$ for the code can be defined as

$$\mathbf{H}(D) = \mathbf{B}_1^T(D). \tag{2.27}$$

The matrix $\mathbf{H}(D)$ has rank $(n-k)$. Note that the parity-check matrix of a code is not unique, as we can easily obtain another parity-check matrix from $\mathbf{H}(D)$ if we premultiply it with any $(n-k) \times (n-k)$ nonsingular matrix. Another algorithm for finding the parity-check matrix $\mathbf{H}(D)$, given the encoder $\mathbf{G}(D)$, is given in [362, Sec. 2.3, 2.4].

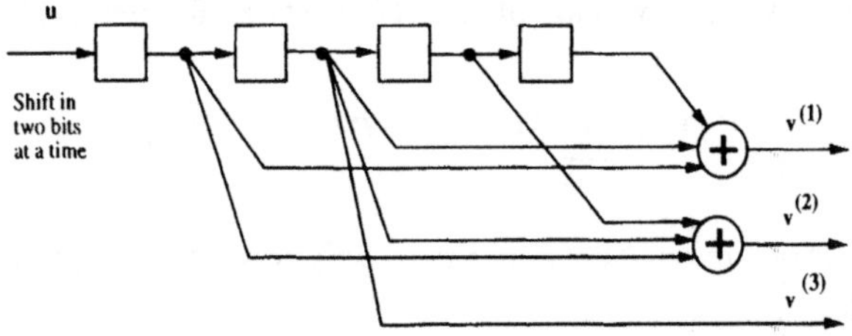


**Figure 2.4** Single shift register implementation of a (3, 2, 1) encoder.

## *Single Shift Register Encoder*

An alternative representation of a rate-$k/n$ encoder is obtained if a single shift register of length $K = \nu + k$ is used, and the inputs are shifted in $k$ bits at a time [100]. This representation will be used in Chapters 4, 10, and 11. A rate-$k/n$ encoder is characterized by $n$ generator polynomials $G^{(j)}(D)$, where[5]

$$G^{(j)}(D) = G_0^{(j)} + G_1^{(j)} D + \cdots + G_{K-1}^{(j)} D^{K-1}, \; j = 1, \ldots, n. \qquad (2.28)$$

In this case, there is a single composite input $\mathbf{u}(D) = \sum_{i=1}^{k} D^{i-1}\, \mathbf{u}^{(i)}(D^k)$, and the encoding operation is

$$\mathbf{v}^{(j)}(D) = \mathbf{u}(D)\, G^{(j)}(D), \;\; j = 1, \ldots, n. \qquad (2.29)$$

The composite encoded output is $\mathbf{v}(D) = \sum_{j=1}^{n} D^{j-1}\, \mathbf{v}^{(j)}(D^n)$. In this notation, we must have $\nu = km$, since the single shift by $k$ positions must preserve the order within the $k$-tuples even if there exist $\nu_i < m$ for some $i = 1, \ldots, k$.

**Example 2.11** Figure 2.4 shows the single shift register implementation of the (3, 2, 1) encoder of Example 2.2. The generator sequences are $G^{(1)} = 1101$, $G^{(2)} = 1110$, and $G^{(3)} = 0100$. The encoder has a single input $\mathbf{u}$ and three outputs $\mathbf{v}^{(j)}$ ($j = 1, 2, 3$). Information block $\mathbf{u}_t = (u_t^{(1)}, u_t^{(2)})$ is shifted in at time $t$ with $u_t^{(2)}$ being shifted in first.

[5]We have chosen upper case letters to represent the generators here to distinguish them from the notation introduced earlier.

Given the matrix $\mathbf{G}(D)$, we can obtain $G^{(j)}(D)$ as follows [207]:

$$G^{(j)}(D) = \sum_{i=1}^{k} D^{i-1}\mathbf{g}_i^{(j)}(D^k), \quad j = 1, \ldots, n. \tag{2.30}$$

On the other hand, given the generator polynomials $G^{(j)}(D)$, we can obtain $\mathbf{g}_i^{(j)}(D)$ as follows:

$$\mathbf{g}_i^{(j)}(D) = G^{(j)}_{[i-1]_k}(D^{1/k}) \; / \; D^{(i-1)/k}, i = 1, \ldots, k, \;\; j = 1, \ldots, n, \tag{2.31}$$

where $[i]_k$ is the congruence class of $i$ (mod $k$), that is the set of integers of the form $i + lk$ for $l = \cdots - 1, 0, 1 \cdots$, and $G^{(j)}_{[i]_k}(D^r)$ is the polynomial obtained from $G^{(j)}(D)$ by selecting only the $[i]_k$ degree terms and then substituting $D^r$ for $D$.

### 2.3.2 Graphical Representations

A better understanding of the convolutional encoding operation is achieved by developing graphical representations. Many useful graphical representations are described in this section.

#### *State Diagram*

A convolutional encoder is a linear sequential circuit and is implemented using shift registers. The contents of the shift registers are called the *state* of the encoder. The state of an $(n, k, m)$ encoder, at time unit $i$, and with an information sequence $\mathbf{u}$ is given by

$$(u^{(1)}_{i-1}, u^{(1)}_{i-2}, \ldots, u^{(1)}_{i-\nu_1}, u^{(2)}_{i-1}, \ldots, u^{(2)}_{i-\nu_2}, \ldots, u^{(k)}_{i-1}, \ldots, u^{(k)}_{i-\nu_k}). \tag{2.32}$$

The next information block $\mathbf{u}_{i+1}$ may change the encoder state. All possible encoder states and the inputs that cause transitions between them can be compactly represented by the *state diagram*. Figure 2.5 shows the state diagram of the (2, 1, 2) encoder of Figure 1.3. Each circle is a state, denoted by the contents of the shift register. The most recent input bit is the least significant bit in the state. The arcs, called branches, denote state transitions caused by an input. Each input produces an output (two bits in this case) which depends upon the input and the current state. The branches in the state diagram are labeled with input/output bits. The encoding operation corresponds to a sequence of state transitions, starting from a known state. In general, each state

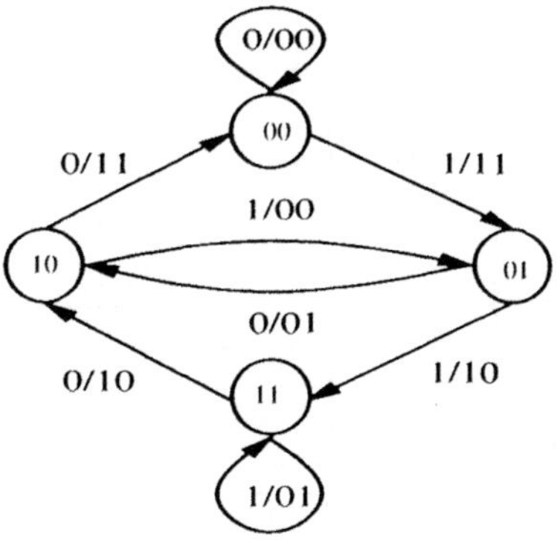


**Figure 2.5** State diagram of a (2, 1, 2) encoder.

has $2^k$ arcs going out from it, corresponding to the information $k$-tuple $\mathbf{u}_i$. For a given information sequence, the corresponding encoded sequence is obtained by traversing the state diagram starting with the all-zero state. Choosing the all-zero state as the starting state is equivalent to initializing all the memory elements of the shift register to zero.

**Example 2.12** *The encoded sequence corresponding to the information sequence* $\mathbf{u} = 1101001$ *is* $\mathbf{v} = 11\ 10\ 10\ 00\ 01\ 11\ 11\ 01\ 11$. *This sequence is the same as that obtained in Example 2.5.*

A sequence of states $\sigma_0 \rightarrow \sigma_1 \rightarrow \cdots \rightarrow \sigma_x$ is called a *cycle* in the state diagram if the starting states $\sigma_0$ is the same as the ending state $\sigma_x$, and all the states in between are distinct. The cycle is said to have length $x$ and weight equal to the sum of the Hamming weights of outputs corresponding to all the branches in the cycle. For instance, the self-loop around the all-zero state in Figure 2.5 is a *zero-weight* cycle of length one. A sequence of states $\delta_0 \rightarrow \delta_1 \rightarrow \cdots \rightarrow \delta_y$ is called a *walk* of length $y$ in the state diagram. It corresponds to a sequence of $y$ branches.

## *Tree Diagram*

A more redundant representation of the encoder is the *tree diagram*, shown in Figure 2.6 for the encoder of Figure 1.3. Each node has two outgoing branches. The upper branch corresponds to an input bit 0, and the lower branch corre-

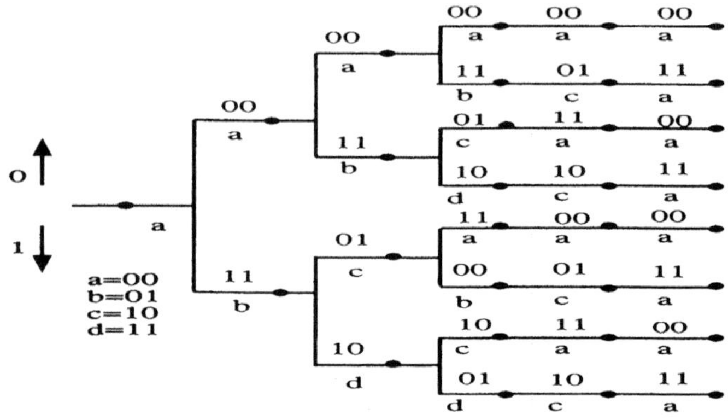


**Figure 2.6** Tree diagram of a (2, 1, 2) encoder.

sponds to an input bit 1. The tree expands from left to right as more input bits are considered. All possible paths are shown explicitly in a tree diagram. The label above each branch is the output. The state of the encoder at each time unit is shown below each branch. The encoder starts in state $a = 00$. After $m = 2$ time units, all four states appear in the tree diagram.

The tree in Figure 2.6 corresponds to a three bit information sequence followed by $m = 2$ terminating 0's. Each path from the root node to a leaf node is a codeword. A codeword obtained in this manner is called a *terminated* codeword. Note that a string of $m$ or more consecutive all-zero information blocks in an information sequence effectively results in two separate terminated codewords. This is because the string of $m$ all-zero information blocks brings the encoder to the initial state, i.e., it "flushes" the memory of the encoder.

## *Trellis Diagram*

A very common representation for the convolutional encoder is the *trellis diagram.* The trellis diagram for the (2, 1, 2) encoder of Figure 1.3 is shown in Figure 2.7. There are four possible states. The starting state is assumed to be $a = 00$. After $m = 2$ time units, the trellis expands to include all four states. Notice the repetition of states in the top and the bottom half of the tree in Figure 2.6, when total number of branches is eight (at time $m + 1 = 3$). This repetition continues as the number of branches in the tree increases. However, it is avoided in the trellis.

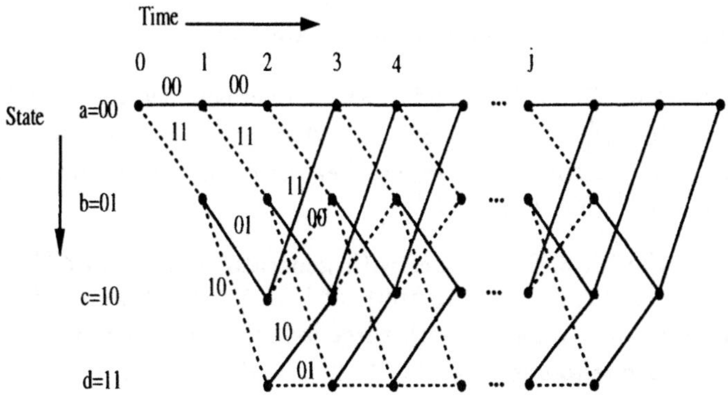


**Figure 2.7** Trellis diagram of a (2, 1, 2) encoder.

Each state in the trellis has two branches leaving it; the dashed branches correspond to input bit 1 and the solid branches correspond to input bit 0. The label on each branch is the corresponding output bits. Last $m$ stages of the trellis are for the termination of the codeword and correspond to an all-zero input. Each codeword is a path starting from state 00 at time $t = 0$ and ending at state 00.

## 2.4 EXERCISES

1. Draw the encoder block diagram of the (3, 1, 3) encoder with $\mathbf{g}^{(1)} = 1101$, $\mathbf{g}^{(2)} = 1010$, and $\mathbf{g}^{(3)} = 1111$. Find its generator matrix $\mathbf{G}$.

2. Find the polynomial generator matrix $\mathbf{G}(D)$ of the (3, 2, 4) encoder with generator sequences $\mathbf{g}_1^{(1)} = 10011$, $\mathbf{g}_1^{(2)} = 10100$, $\mathbf{g}_1^{(3)} = 11111$, $\mathbf{g}_2^{(1)} = 10010$, $\mathbf{g}_2^{(2)} = 00100$, and $\mathbf{g}_2^{(3)} = 01110$. Find its input constraint length, its overall constraint length, its memory order, and its output constraint length.

3. Draw the trellis diagram of the (3, 1, 3) encoder with generator sequences $\mathbf{g}^{(1)} = 1101$, $\mathbf{g}^{(2)} = 1010$, and $\mathbf{g}^{(3)} = 1111$. Find the codeword $\mathbf{v}$ corresponding to the information sequence $\mathbf{u} = 1000$.

4. Draw the encoder block diagram and the state diagram of the (2, 1, 3) encoder with generator polynomials $\mathbf{g}^{(1)}(D) = 1 + D^2 + D^3$ and $\mathbf{g}^{(2)}(D) =$

$1+D+D^2+D^3$. Find the shortest path starting from and ending at state 000 other than the self-loop around state 000.

5. Consider the (3, 2, 4) encoder with generator sequences $\mathbf{g}_1^{(1)} = 10011$, $\mathbf{g}_1^{(2)} = 10100$, $\mathbf{g}_1^{(3)} = 11111$, $\mathbf{g}_2^{(1)} = 10010$, $\mathbf{g}_2^{(2)} = 00100$, and $\mathbf{g}_2^{(3)} = 01110$. Draw its single shift register encoder diagram. Determine the generator sequences $G^{(1)}$, $G^{(2)}$, and $G^{(3)}$.

6. Consider the (5, 3, 2) encoder with generator sequences $G^{(1)} = 101\ 1101$, $G^{(2)} = 011\ 0111$, $G^{(3)} = 110\ 1001$, $G^{(4)} = 010\ 1010$, and $G^{(5)} = 111\ 1101$. Draw its encoder diagram with three inputs and 5 outputs. Determine the generator sequences $\mathbf{g}_i^{(j)}$ for $i = 1, 2, 3$ and $j = 1, \ldots, 5$.

7. Consider the (3, 2, 3) encoder with $\mathbf{g}_1^{(1)} = 10011$, $\mathbf{g}_1^{(2)} = 10100$, $\mathbf{g}_1^{(3)} = 11111$, $\mathbf{g}_2^{(1)} = 10010$, $\mathbf{g}_2^{(2)} = 00100$, and $\mathbf{g}_2^{(3)} = 01110$. Find its generator polynomials $\mathbf{g}_i^{(j)}(D)$ for $i = 1, 2$ and $j = 1, 2, 3$. Draw its encoder block diagram. For $\mathbf{u}^{(1)}(D) = 1 + D + D^2$ and $\mathbf{u}^{(2)}(D) = D + D^3$, find the encoded outputs $\mathbf{v}^{(1)}(D)$, $\mathbf{v}^{(2)}(D)$, and $\mathbf{v}^{(3)}(D)$.

8. Draw the state and trellis diagrams of the (2, 1, 1) encoder with generator sequences $\mathbf{g}^{(1)} = 10$ and $\mathbf{g}^{(2)} = 11$.

9. Draw the encoder block diagram and the state diagram of the (2, 1, 2) encoder with generator sequences $\mathbf{g}^{(1)} = 100$ and $\mathbf{g}^{(2)} = 001$.

10. Write a computer program to implement encoding and syndrome generation using the (2, 1, 5) encoder with generator sequences $\mathbf{g}^{(1)} = 10011$ and $\mathbf{g}^{(2)} = 11101$. For encoding, your program should accept an information sequence of a user specified length as input and produce an encoded sequence as output. The encoding should begin and end with encoder in the all-zero state. For syndrome generation, your program should accept a received sequence of a user specified length as input and produce the corresponding syndrome sequence as output.

# 3

# DISTANCE PROPERTIES OF CONVOLUTIONAL CODES AND ENCODERS

Definitions and representations of convolutional codes and encoders were discussed in Chapter 2. Our objective in this chapter is to study their distance structure. The distance properties of a code or an encoder determine its error-correcting and detecting capabilities and are useful in analyzing the performance of various decoding techniques. Many of the distance parameters discussed in this chapter are also useful in selecting good encoders.

## 3.1 A DISTANCE METRIC

Consider the two-dimensional Euclidean space $\mathcal{R}^2$. Each point $\mathbf{v}$ (called a *vector*) in the space is specified by an ordered pair of real numbers $(x, y)$, called the abscissa and the ordinate, respectively. The point $\mathbf{0} = (0, 0)$ is called the *origin*. The length of a vector $\mathbf{v} = (x, y) \in \mathcal{R}^2$ is given by

$$l_E(\mathbf{v}) = \sqrt{x^2 + y^2}. \tag{3.1}$$

The difference of two vectors $\mathbf{v}_1 = (x_1, y_1)$ and $\mathbf{v}_2 = (x_2, y_2)$ is also a vector in $\mathcal{R}^2$, namely, $\mathbf{v}_1 - \mathbf{v}_2 = (x_1 - x_2, y_1 - y_2)$. Given two vectors $\mathbf{v}_1 = (x_1, y_1)$ and $\mathbf{v}_2 = (x_2, y_2)$, the Euclidean distance between them is

$$d_E(\mathbf{v}_1, \mathbf{v}_2) = \sqrt{(x_1 - x_2)^2 + (y_1 - y_2)^2} = l_E(\mathbf{v}_1 - \mathbf{v}_2). \tag{3.2}$$

From (3.1) and (3.2), it is clear that the length of a vector $\mathbf{v}$ can also be defined as the distance between $\mathbf{v}$ and the origin $\mathbf{0}$, i.e., $l_E(\mathbf{v}) = d_E(\mathbf{v}, \mathbf{0})$. The distance measure so defined is a *metric* for $\mathcal{R}^2$ since it satisfies the following properties:

1. $d_E(\mathbf{u}, \mathbf{v}) > 0$ for $\mathbf{u} \neq \mathbf{v}$, and $d_E(\mathbf{u}, \mathbf{v}) = 0$ for $\mathbf{u} = \mathbf{v}$,

2. $d_E(\mathbf{u}, \mathbf{v}) = d_E(\mathbf{v}, \mathbf{u})$,

3. $d_E(\mathbf{u}, \mathbf{v}) + d_E(\mathbf{v}, \mathbf{w}) \geq d_E(\mathbf{u}, \mathbf{w})$.

These conditions are called positive definiteness, symmetry, and triangle inequality, respectively.

If a code is defined as a subset of $\mathcal{R}^2$, then the Euclidean distance defined above can be used to characterize the error-correcting capability of the code. Suppose that a codeword $\mathbf{v}$, transmitted over a noisy channel, is received as the vector $\mathbf{r} \in \mathcal{R}^2$. Let the error $\mathbf{e} \in \mathcal{R}^2$ introduced by the channel be such that $\mathbf{r} = \mathbf{v} + \mathbf{e}$. The length $l_E(\mathbf{e})$ is a measure of the magnitude of noise on the channel. A nearest-neighbor decoding algorithm would decode the received vector $\mathbf{r}$ as the codeword $\mathbf{v}$ that is closest in distance to $\mathbf{r}$. A decoding error occurs while using this rule if the error is large enough to make $\mathbf{r}$ closer to a codeword $\mathbf{w} \neq \mathbf{v}$. We can make this notion more precise by defining a distance parameter that characterizes a code $C$. This parameter is called the minimum distance of the code and is given by

$$d_{min} = min \; \{d_E(\mathbf{v}_i, \mathbf{v}_j), \; \mathbf{v}_i \neq \mathbf{v}_j \in C\}. \tag{3.3}$$

Thus, a decoding error using nearest neighbor decoding will not occur if $l_E(\mathbf{e}) < d_{min}/2$. The concept of distance can be generalized to the $n$-dimensional Euclidean space $\mathcal{R}^n$. The error-correcting capability of a code defined over $\mathcal{R}^n$ can be accordingly determined.

Now, let us consider the set $\{GF(2)\}^n$ of binary $n$-tuples which is an $n$-dimensional vector space over $GF(2)$. Any vector in this space is given by $\mathbf{v} = (v_1, v_2, \ldots, v_n)$, where $v_i \in GF(2)$ for $i = 1, \ldots, n$. We are interested in measures of length and distance for vectors in $\{GF(2)\}^n$. A measure similar to the length of a vector in $\mathcal{R}^n$ is called the *Hamming weight.*

**Definition 3.1** *The Hamming weight $w_H(\mathbf{v})$ of a vector $\mathbf{v} \in \{GF(2)\}^n$ is defined to be the number of nonzero elements in the $n$-tuple.*

For example, the Hamming weight of the 5-tuple 11101 is 4, and the Hamming weight of the all-zero 5-tuple 00000 is 0.

A distance measure similar to the Euclidean distance given above and called the *Hamming distance* can be defined for vectors in $\{GF(2)\}^n$.

**Definition 3.2** *Let* $\mathbf{v}_1, \mathbf{v}_2 \in \{GF(2)\}^n$. *The Hamming distance* $d_H(\mathbf{v}_1, \mathbf{v}_2)$ *between* $\mathbf{v}_1$ *and* $\mathbf{v}_2$ *is given by*

$$d_H(\mathbf{v}_1, \mathbf{v}_2) = w_H(\mathbf{v}_1 - \mathbf{v}_2). \tag{3.4}$$

Note that subtraction is the same as addition over $GF(2)$. For instance, the Hamming distance between 11101 and 11001 is 1, and the the Hamming distance between 10101 and 01010 is 5.

The Hamming weight of a vector $\mathbf{v}$ is also the Hamming distance between $\mathbf{v}$ and the all-zero $n$-tuple $\mathbf{0}$, i.e., $w_H(\mathbf{v}) = d_H(\mathbf{v}, \mathbf{0})$. It is easy to verify that the Hamming distance is a metric for the vector space $\{GF(2)\}^n$ since it satisfies the three conditions given above. Finally, the definitions of Hamming weight and Hamming distance given here for vectors in $\{GF(2)\}^n$ can be generalized to $n$-tuples over $\{GF(q)\}^n$.

Although in practice we always deal with finite length sequences of $n$-tuples $\mathbf{v}_t^{(j)} \in \{GF(q)\}^n$, we have to allow for the possibility, in principle, of infinite sequences. The Hamming weight $w_H(\mathbf{v})$ of an infinite sequence $\mathbf{v}$ of $n$-tuples $\mathbf{v}_t^{(j)}$ is defined as the sum of the Hamming weights of all of its components. This weight is finite if only a finite number of $n$-tuples $\mathbf{v}_t^{(j)}$ have nonzero weights; otherwise its infinite. The Hamming distance between two infinite sequences $\mathbf{v}$ and $\mathbf{w}$ is $d_H(\mathbf{v}, \mathbf{w}) = w_H(\mathbf{v} - \mathbf{w})$, which also may be finite or infinite.

## 3.2 DISTANCE MEASURES

With this background, we are now ready to define various distance measures of convolutional codes and encoders.

### 3.2.1 Column Distance Function

The column distance function of a convolutional encoder is an important distance measure [91, 92]. Many other distance measures are defined based on the column distance function and are discussed later in this chapter.

Let $\mathbf{u}_{[0,j]} = (\mathbf{u}_0, \mathbf{u}_1, \ldots, \mathbf{u}_j)$ represent the first $(j+1)$ information blocks and $\mathbf{v}_{[0,j]} = (\mathbf{v}_0, \mathbf{v}_1, \ldots, \mathbf{v}_j)$ represent the first $(j+1)$ encoded blocks. Further, let

$[\mathbf{G}]_j$ contains the first $n(j+1)$ columns of $\mathbf{G}$, i.e.,

$$[\mathbf{G}]_j = \begin{bmatrix} \mathbf{G}_0 & \mathbf{G}_1 & \cdots & \mathbf{G}_j \\ & \mathbf{G}_0 & \cdots & \mathbf{G}_{j-1} \\ & & \ddots & \vdots \\ & & & \mathbf{G}_0 \end{bmatrix}, \tag{3.5}$$

for $j \leq m$, or

$$[\mathbf{G}]_j = \begin{bmatrix} \mathbf{G}_0 & \mathbf{G}_1 & \cdots & \mathbf{G}_m & & & & & \\ & \mathbf{G}_0 & \cdots & \mathbf{G}_{m-1} & \mathbf{G}_m & & & & \\ & & \ddots & \vdots & \mathbf{G}_{m-1} & \ddots & & & \\ & & & \ddots & \vdots & & \ddots & & \\ & & & & \mathbf{G}_1 & \cdots & \cdots & \mathbf{G}_m & \\ & & & & \mathbf{G}_0 & \mathbf{G}_1 & \cdots & \mathbf{G}_{m-1} & \mathbf{G}_m \\ & & & & & \mathbf{G}_0 & \cdots & \mathbf{G}_{m-2} & \mathbf{G}_{m-1} \\ & & & & & & \ddots & \vdots & \vdots \\ & & & & & & & \mathbf{G}_0 & \mathbf{G}_1 \\ & & & & & & & & \mathbf{G}_0 \end{bmatrix}, \tag{3.6}$$

for $j > m$. Then, the encoding operation truncated after $(j+1)$ time units is given by

$$\mathbf{v}_{[0,j]} = \mathbf{u}_{[0,j]}[\mathbf{G}]_j. \tag{3.7}$$

**Definition 3.3** *The column distance function of order $j$ of a convolutional encoder is defined as*

$$\begin{aligned} d_j &= \min\{d_H(\mathbf{u}_{[0,j]}[\mathbf{G}]_j, \mathbf{u}'_{[0,j]}[\mathbf{G}]_j) : \mathbf{u}_0 \neq \mathbf{u}'_0\} \\ &= \min\{w_H(\mathbf{u}_{[0,j]}[\mathbf{G}]_j) : \mathbf{u}_0 \neq \mathbf{0}\}. \end{aligned} \tag{3.8}$$

In (3.8), linearity of convolutional codes is used to define the column distance function as the weight of the minimum weight encoded subsequence resulting from an information subsequence with a nonzero first block. Note that the column distance function is monotonically nondecreasing (see exercises).

It is easy to compute the column distance function of a given order of an encoder from its tree diagram. Considering the root node to be at level $j = 0$, the node(s) at level $j = 1$ corresponding to nonzero $\mathbf{u}_0$ can be labeled with the branch weight, i.e. the Hamming weight of the output $\mathbf{v}_0$ corresponding to the

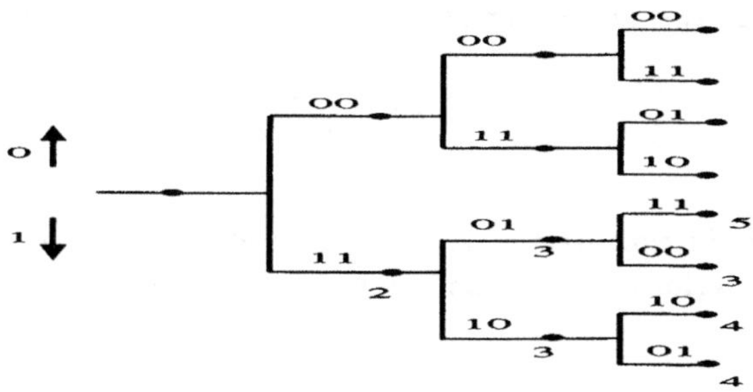


**Figure 3.1** Tree diagram of a (2, 1, 2) encoder.

branch. The minimum among these weights is then $d_0$, the column distance function of order $j = 0$. For the next and all the following levels, the labels on all the nodes that can be reached from the nodes labeled in the previous level are the sum of the weights of starting nodes and the branch weight of the current branch. Again, minimum of the labels at level $j+1$ is equal to $d_j$.

**Example 3.1** Consider the tree diagram of the (2, 1, 2) encoder with generator sequences $\mathbf{g}^{(1)} = 101$ and $\mathbf{g}^{(2)} = 111$, shown in Figure 3.1 (only $j = 3$ levels are shown). Only one node can be reached by taking the branch corresponding to $u_0 = 1$, with branch weight 2. That node is labeled 2 in the figure, and we have $d_0 = 2$. At the next level, two nodes can be reached, corresponding to $u_1 = 0$ and $u_1 = 1$, respectively. Each of these branches has weight equal to 1, and the nodes at level $j = 2$ are labeled with weights equal to 3. Hence, $d_1 = 3$. The labels at the next levels can be computed similarly and are found to be 5, 3, 4, and 4. Hence, $d_2 = 3$, corresponding to $\mathbf{u}_{[0,2]} = 101$. Carrying on in this manner, the column distance function $d_j$ of this encoder, for orders $j = 0, \ldots, 9$, is $d_0 = 2$, $d_1 = d_2 = 3$, $d_3 = d_4 = 4$, $d_5 = d_6 = d_7 = d_8 = d_9 = 5$. Note that the column distance function seems to stabilize at a value of 5. We discuss this in the next section.

The column distance function is useful in determining the error-correcting capability of the *truncated* encoder $[\mathbf{G}]_j$ given by (3.5) and (3.6). If a decoder bases its decision on $(j+1)$ blocks, then any subsequence $\mathbf{w}_{[0,j]}$ of length $(j+1)$ $n$-tuples that is at a distance $\lfloor (d_j - 1)/2 \rfloor$ or less from $\mathbf{v}_{[0,j]}$ is closer to $\mathbf{v}_{[0,j]}$ than any other code subsequence.

**Definition 3.4** *The error-correcting capability* $t_j$ *of a truncated encoder* $[\mathbf{G}]_j$ *is*

$$t_j = \lfloor \frac{d_j - 1}{2} \rfloor, \tag{3.9}$$

where $\lfloor x \rfloor$ is the integer part of $x$. The parameter $t_j$ will be used in Chapter 8 to analyze the performance of table look-up decoding.

### 3.2.2 Free Distance

It is interesting to consider the distance between any two codewords in a convolutional code. A distance measure called *free distance*, denoted by $d_{free}$, was suggested by Costello [91, 92].[1] Let $\mathbf{u}$ and $\mathbf{u}'$ be two information sequences, and let $\mathbf{v}$ and $\mathbf{v}'$ be terminated codewords corresponding to $\mathbf{u}$ and $\mathbf{u}'$, respectively.

**Definition 3.5** *The free distance of a convolutional code is defined as*

$$\begin{aligned} d_{free} &= min\ \{d_H(\mathbf{v},\ \mathbf{v}') : \ \mathbf{u} \neq \mathbf{u}'\} \\ &= min\ \{w_H(\mathbf{v}) : \ \mathbf{u} \neq \mathbf{0}\} \\ &= min\ \{w_H(\mathbf{uG}) : \ \mathbf{u} \neq \mathbf{0}\}. \end{aligned} \tag{3.10}$$

Thus, the free distance is the minimum Hamming distance between any two distinct codewords. Again, by linearity of convolutional codes, the free distance is also the weight of the minimum weight nonzero codeword. The free distance is a parameter of the convolutional code, while the column distance function is a parameter of the convolutional encoder. Of course, the free distance is also a parameter of any encoder that generates the code.

**Example 3.2** Consider the (2, 1, 2) encoder with generator sequences $\mathbf{g}^{(1)} = 101$ and $\mathbf{g}^{(2)} = 111$. Its free distance is $d_{free} = 5$. Note that free distance of the code generated by this encoder is also 5.

How are the column distance function $d_j$ and the free distance $d_{free}$ of an encoder related? It seems possible that $d_j$ will be equal to $d_{free}$ for some $j$. Consider the value of $d_j$ as $j \to \infty$, denoted by $d_\infty$. Is $d_{free} = d_\infty$? This is not always the case, as illustrated by the following example.

[1] The term "free distance" was coined by Massey [315].

**Example 3.3** The (2, 1, 2) encoder with generator sequences $\mathbf{g}^{(1)} = 101$ and $\mathbf{g}^{(2)} = 110$ has free distance $d_{free} = 4$, achieved by the terminated codeword corresponding to information sequence $\mathbf{u} = 100$. However, an information sequence of infinite length consisting of all 1's generates an encoded subsequence with weight 3. Hence, $d_\infty = 3 \neq d_{free}$.

Encoders like the one in Example 3.3 that generate a finite weight codeword for an infinite weight information sequence were named *catastrophic* by Massey [225]. For a catastrophic encoder, it is possible to have $d_\infty < d_{free}$. The free distance of a noncatastrophic encoder can be defined as $d_{free} = d_\infty = \lim_{j\to\infty} d_j$. Clearly, it is impractical to compute this limit. Fortunately, there is a finite $j_0$ such that $d_j = d_{free}$ for $j \geq j_0$, i.e., a finite length information sequence achieves the free sequence.

An encoder is an *optimum free distance* (OFD) encoder if its free distance is equal or superior to that of any other encoder of the same rate and constraint length [221]. Many of the OFD encoders are listed in Section 4.12.

The *maximum error-correcting capability* of a code or an encoder is determined by its free distance.

**Definition 3.6** *The maximum error-correcting capability $t_{free}$ of a code or an encoder is*

$$t_{free} = \lfloor \frac{d_{free} - 1}{2} \rfloor. \tag{3.11}$$

In Chapter 5, $t_{free}$ is used to analyze the performance of Viterbi decoding.

## 3.2.3 Minimum Distance

The column distance function of order $j = m$, where $m$ is the memory order of the encoder, is called its *minimum distance* [474]. It is defined as follows:

**Definition 3.7** *The minimum distance of a convolutional encoder is $d_{min} = d_m$.*

The minimum distance is the minimum weight of one output constraint length ($n_A = n(m+1)$ symbols) subsequence of a codeword with a nonzero first information block.

**Example 3.4** The column distance function of the (2, 1, 2) encoder with generator sequences $\mathbf{g}^{(1)} = 101$ and $\mathbf{g}^{(2)} = 111$ was evaluated in Example 3.1. From Figure 3.1, we see that at level $j = m = 2$, four code sub-sequences have weights 5, 3, 4, and 4, and $d_2 = 3$. Therefore, this encoder has minimum distance $d_{min} = d_2 = 3$.

An encoder is an *optimum minimum distance* (OMD) encoder if its minimum distance is equal or superior to that of any other encoder of the same rate and constraint length [221]. Listings of OMD encoders are included in Section 4.12.

The minimum distance has traditionally been used in evaluating the performance of decoding schemes that base the decoding of an information block $\mathbf{u}_t$ on one output constraint length of the received symbols $\mathbf{r}_{[t,t+m]}$. A common example of such a decoding scheme is feedback decoding which will be discussed in Chapters 7 and 8. The *feedback error-correcting capability* of an encoder is determined by its minimum distance.

**Definition 3.8** *The feedback error-correcting capability $t_{min}$ of a truncated encoder $[\mathbf{G}]_m$ is*

$$t_{min} = \lfloor \frac{d_{min} - 1}{2} \rfloor. \tag{3.12}$$

## 3.2.4 Distance Profile

Massey and Costello have observed that good computational performance with sequential decoding is achievable if the column distance function "grows as rapidly as possible" [318]. Motivated by the need to capture the growth of the column distance function as its order is increased, Johannesson defined a $(\nu + 1)$-tuple called the *distance profile* [221].

**Definition 3.9** *The distance profile of a convolutional encoder is given by*

$$\mathbf{d} = [d_0, d_1, ..., d_m]. \tag{3.13}$$

A distance profile $\mathbf{d}$ is said to be superior to a distance profile $\mathbf{d}'$ if there is some $j_0$ such that

$$d_j \begin{cases} = d'_j, & j = 0, 1, ..., j_0 - 1 \\ > d'_j, & j = j_0 \end{cases} \tag{3.14}$$

Thus, $\mathbf{d} > \mathbf{d}'$ indicates that $d_j$ grows faster with $j$ than $d'_j$ initially, although it could be the case that $d'_j > d_j$ for a sufficiently large $j$.

**Example 3.5** The distance profile of the (2, 1, 2) encoder with generator sequences $\mathbf{g}^{(1)} = 101$ and $\mathbf{g}^{(2)} = 111$ is $\mathbf{d} = [2, 3, 3]$. Another (2, 1, 2) encoder with generator sequences $\mathbf{g}^{(1)} = 100$ and $\mathbf{g}^{(2)} = 011$ has the distance profile $\mathbf{d}' = [1, 2, 3]$. It is clear that $\mathbf{d}$ is superior to $\mathbf{d}'$.

An encoder is an *optimum distance profile* (ODP) encoder if its distance profile is equal or superior to that of any other encoder of the same rate and constraint length [221]. Listings of ODP encoders are given in Section 4.12.

The distance profile of an encoder does not capture the growth of the column distance function beyond the minimum distance. For this purpose, we can use the *generalized distance profile* [224].

**Definition 3.10** *The generalized distance profile* $\mathbf{d}_l$ *of a convolutional encoder, for some integer* $l \geq 0$, *is given by*

$$\mathbf{d}_l = [d_0, d_1, d_2 ..., d_l]. \tag{3.15}$$

### 3.2.5 Distance Spectrum

The number of codewords of a specific weight generated by an encoder is useful in analyzing the performance of Viterbi decoding. Let $n(d_{free} + i)$ denote the number of weight $d_{free} + i$ codewords that correspond to a nonzero first information block and all-zero last $m$ information blocks, and let it be called the $(i+1)$th *spectral component* [70]. An enumeration of the number of codewords of specific weights is called *distance spectrum* of the encoder.

**Definition 3.11** *The distance spectrum of a code or an encoder is the sequence*

$$n(d_{free} + i), \quad i = 0, 1, 2, \ldots. \tag{3.16}$$

A related parameter is the total number of nonzero information bits, $n_b(d_{free} + i)$, in all weight $d_{free} + i$ codewords. It is useful in calculating a bound on the bit error probability when Viterbi decoding is used.

**Example 3.6** Consider the code generated by the (2, 1, 2) encoder with generator sequences $\mathbf{g}^{(1)} = 101$ and $\mathbf{g}^{(2)} = 111$. The free distance of this encoder is $d_{free} = 5$, and the first 10 components of its distance spectrum are {1, 2, 4, 8, 16, 32, 64, 128, 256, 512}. This means that there is only one codeword with weight $d_{free} = 5$, two codewords with weight $d_{free} + 1 = 6$, etc.

Since only terminated codewords are counted, the distance spectrum is a parameter of the code and of the encoders that generate it. A fast algorithm to compute the distance spectrum of convolutional codes is given by Cedervall and Johannesson [70].

## 3.2.6 Decision Depth

The column distance function $d_j$ of order $j$ is the weight of the minimum weight code subsequence of length $(j+1)$ blocks resulting from an information subsequence with a nonzero first block ($\mathbf{u}_0 \neq 0$). It is a measure of the "spread" of code subsequences of length $(j+1)$ blocks. The inverse of this measure is the answer to the following question: given a weight $w$, what is the number of blocks such that all unmerged code subsequences with nonzero first information block have weight greater than $w$? By unmerged code subsequences we mean those paths in the trellis that have not merged with the all-zero path. Such a question is relevant when one wants to know the lengths of code subsequences that one needs to examine to achieve a guaranteed "spread". That is, if one wants a nearest-neighbor decoding algorithm to guarantee correction of $t$ or fewer errors, a "spread" greater than $2t$ is required. An encoder parameter that provides an answer to this question is the *decision depth*, defined by Anderson and Balachandran [15].

**Definition 3.12** *The decision depth function $L_D(w)$ for a convolutional encoder is the first trellis depth at which all incorrect subset paths, except those merged to the transmitted path, exceed weight $w$.*

The decision depth is useful in selecting the length of path memory needed by a convolutional decoder. For listings of decision depths of encoders of rate one-third, one-half, two-thirds, and three-fourths, see [15, 13].

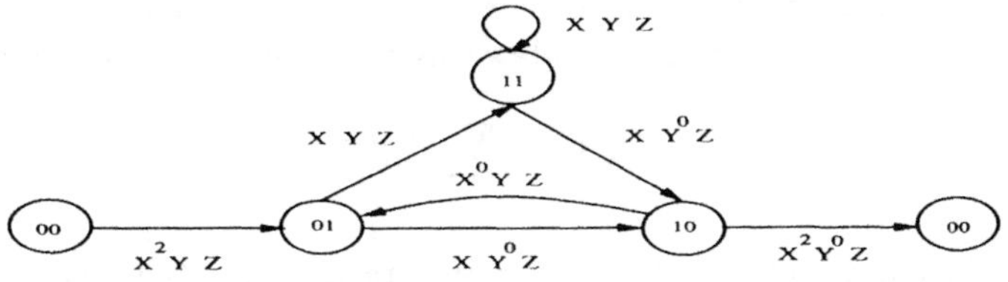


**Figure 3.2** Modified state diagram of a (2, 1, 2) encoder.

# 3.3 ALGORITHMS FOR COMPUTING DISTANCE MEASURES

First, we present a method based on the transfer function of a convolutional encoder. The methods based on the decoding algorithms are then summarized.

## 3.3.1 Transfer Function of an Encoder

The transfer function of an encoder is useful in evaluating its weight distribution. The method is based on a modified encoder state diagram obtained by splitting the all-zero state into an initial and a final all-zero state.

**Example 3.7** Consider the (2, 1, 2) encoder shown in Figure 1.3. Let the starting point and the ending point of any walk through the state diagram be the all-zero state. The all-zero state is split into an initial and a final state. The resulting modified state diagram is shown in Figure 3.2.

The branches of the modified state diagram are labeled as follows. The exponent of the letter $X$ indicates the weight of the encoded branch, that of the letter $Y$ indicates the weight of the information branch, and finally, the exponent of the letter $Z$ indicates the length of the path in branches.

The generating function of the encoder is given by

$$T(X, Y, Z) = \sum_{i,j,l} A_{i,j,l} X^i Y^j Z^l, \tag{3.17}$$

where $A_{i,j,l}$ is the number of codewords with weight $i$, whose associated information sequence has weight $j$, and whose length is $l$ branches. A simplified transfer function is obtained by considering only the weight of the codewords. In this case, $A_i$ is the number of codewords with weight $i$, and we get

$$T(X) = \sum_i A_i X^i. \tag{3.18}$$

For a description of transfer function computation, see [454, 285].

**Example 3.8** The state diagram shown in Figure 3.2 is also a signal flow graph, with the state equations representing the different states given by

$$\begin{aligned}
\psi_{01} &= X^2YZ\psi_{00_i} + YZ\psi_{10} \\
\psi_{10} &= XZ\psi_{01} + XZ\psi_{11} \\
\psi_{11} &= XYZ\psi_{01} + XYZ\psi_{11} \\
\psi_{00_f} &= X^2Z\psi_{10},
\end{aligned} \tag{3.19}$$

where $\psi_x$'s are dummy variables for partial paths to the intermediate nodes, i.e., they represent partial weights and lengths associated with the state $x$. The initial and final all-zero states are $00_i$ and $00_f$, respectively. The transfer function $T(X,\ Y,\ Z) = \psi_{00_f}/\psi_{00_i}$, obtained by computing a simultaneous solution of (3.19), is

$$T(X,\ Y,\ Z) = \frac{X^5Y^3Z}{1 - XYZ(1+Y)}. \tag{3.20}$$

The simplified transfer function, obtained by setting $Y = Z = 1$, is

$$T(X) = \frac{X^5}{1-2X} = X^5 + 2X^6 + 4X^7 + \cdots + 2^l X^{l+5} + \cdots. \tag{3.21}$$

This expansion shows that there is one codeword with weight 5, two codewords with weight 6, etc, confirming the results in Example 3.6.

### 3.3.2 Algorithms Based on Convolutional Decoding Techniques

Many researchers have proposed the use of convolutional decoding algorithms in computing various distance measures of an encoder. In convolutional decoding, the received sequence is compared block by block to all possible encoded sequences, and the encoded sequence closest in Hamming distance to the received sequence is selected as the estimate of the transmitted encoded sequence.

The encoder tree and trellis are structures that allow easy enumeration of all the encoded sequences and are often used in decoding schemes. The Viterbi decoding described in Chapter 5 uses the encoder trellis, while sequential decoding described in Chapter 6 uses the encoder tree. It is easy to modify these schemes to evaluate various distance parameters.

Forney [148] has described the use of a sequential decoding simulator in determining the distance spectrum. Bahl *et al.* [22] have given a bi-directional algorithm for computing the free distance. This algorithm was modified by Larsen [260]. Chevillat [78] has given an algorithm based on sequential decoding for computing the column distance function. Cedervall and Johannesson [70] have given a fast algorithm for computing the distance spectrum. Their algorithm uses the distance profiles of the encoder and its reverse (obtained by reversing all the generators) to limit the search considerably. Another algorithm was reported in [174]. The Viterbi algorithm can also be modified to evaluate the column distance function and the free distance.

## 3.4 DISTANCE BOUNDS

We summarize known bounds on the free distance $d_{free}$ of convolutional codes [93]. A Gilbert-like lower bound is the following:

$$\frac{d_{free}}{n_A} \geq H^{-1}(1 - R), \tag{3.22}$$

where $R = k/n$ is the coding rate, $n_A = n(m+1)$ is the output constraint length, and $H(x)$ is the entropy function for a binary alphabet, given by

$$H(x) = -x \, log_2 \, x - (1-x) \, log_2 \, (1-x), \;\; 0 \leq x \leq 1. \tag{3.23}$$

Heller [195] has given the following upper bound on the free distance $d_{free}$ of a rate-$1/n$ convolutional code:

$$d_{free} \leq min_{j \geq 1} \lfloor \frac{n}{2} \frac{2^j}{2^j - 1}(\nu + j) \rfloor. \tag{3.24}$$

## 3.5 EXERCISES

1. Prove that the column distance function of a convolutional encoder is a monotonically nondecreasing function.

2. Compute the free distance $d_{free}$ and the column distance function of order $j = 0, \ldots, 5$ of the (2, 1, 2) encoder with generator sequences $\mathbf{g}^{(1)} = 110$ and $\mathbf{g}^{(2)} = 011$. Comment on $\lim_{j\to\infty} d_j$ and compare it with $d_{free}$.

3. Calculate the distance profile $\mathbf{d}$ of the (3, 1, 3) encoder with generator sequences $\mathbf{g}^{(1)} = 1101$, $\mathbf{g}^{(2)} = 1010$, and $\mathbf{g}^{(3)} = 1111$. Find its minimum distance $d_{min}$.

4. Draw augmented state diagram for the (2, 1, 3) encoder with generator sequences $\mathbf{g}^{(1)} = 1011$ and $\mathbf{g}^{(2)} = 1111$.

5. Draw the augmented state diagram for the (2, 1, 1) encoder with generator sequences $\mathbf{g}^{(1)} = 10$ and $\mathbf{g}^{(2)} = 11$, and find its transfer function $T(X)$.

6. The transfer function of the (2, 1, 3) encoder with generator sequences $\mathbf{g}^{(1)} = 1011$ and $\mathbf{g}^{(2)} = 1111$ is

$$T(X) = \frac{X^6 + X^7 - X^8}{1 - 2X - X^3}.$$

   Using long division, find its free distance $d_{free}$. How many codewords have weight equal to 8?

7. Determine the (2, 1, 7) encoder with the best generalized distance profile among the following:

   - $\mathbf{d}_{l_A} = [2\ 3\ 3\ 4\ 4\ 5\ 5\ 5\ 5\ 6\ 6\ 6\ 6\ 7\ 7\ 8\ 8\ 8\ 9\ 9\ 9\ 10\ 10\ 10\ \cdots]$.
   - $\mathbf{d}_{l_B} = [2\ 3\ 3\ 4\ 4\ 4\ 4\ 5\ 5\ 5\ 5\ 5\ 5\ 5\ 6\ 6\ 6\ 7\ 7\ 8\ 8\ 8\ 8\ 8\ \cdots]$.
   - $\mathbf{d}_{l_C} = [2\ 3\ 3\ 4\ 4\ 5\ 5\ 5\ 6\ 6\ 7\ 7\ 7\ 8\ 8\ 9\ \cdots]$.

   Identify the encoder(s) with best $t_{min}$.

8. In Section 3.2.1, a method based on the encoder tree diagram for computing the column distance function was given. Develop a similar method based on the encoder trellis diagram, and use it to determine the column distance function $d_j$ of order $j = 0, \ldots, 5$ for the (2, 1, 2) encoder with generator sequences $\mathbf{g}^{(1)} = 100$ and $\mathbf{g}^{(2)} = 101$.

9. Write a computer program to implement the tree-based and trellis-based algorithms to find the distance spectrum. Compare the number of computations performed. Suggest methods for reducing this number (Hint: see [70]).

# 4

# CLASSES OF CONVOLUTIONAL CODES AND ENCODERS

The codes and encoders were analyzed according to their distance properties in Chapter 3. In this chapter, we will discuss classes of convolutional codes and encoders with the objective of further understanding their structure.

## 4.1 FIXED AND TIME-VARYING ENCODERS

In Chapter 2, convolutional encoding was represented using a semi-infinite generator matrix $\mathbf{G}$ by (2.9) - (2.11). This representation can be generalized as follows:

$$\mathbf{v}_t = \mathbf{u}_t\mathbf{G}_0(t) + \mathbf{u}_{t-1}\mathbf{G}_1(t) + \cdots + \mathbf{u}_{t-m}\mathbf{G}_m(t), \qquad (4.1)$$

where each $\mathbf{G}_i(t)$ is a $k \times n$ matrix over $F$, representing a mapping from $F^{(m+1)k}$ to $F^n$ [315]. Such an encoder is called a *time-varying* convolutional encoder. The matrices $\mathbf{G}_i(t)$ of a *fixed* convolutional encoder do not depend on $t$, so we can simply write, as we have in (2.11), $\mathbf{v}_t = \mathbf{u}_t\mathbf{G}_0 + \mathbf{u}_{t-1}\mathbf{G}_1 + \cdots + \mathbf{u}_{t-m}\mathbf{G}_m$.

## 4.2 SYSTEMATIC AND NONSYSTEMATIC ENCODERS

Convolutional encoders are commonly classified based on whether the $k$ information sequences appear intact as part of the $n$ encoded sequences.

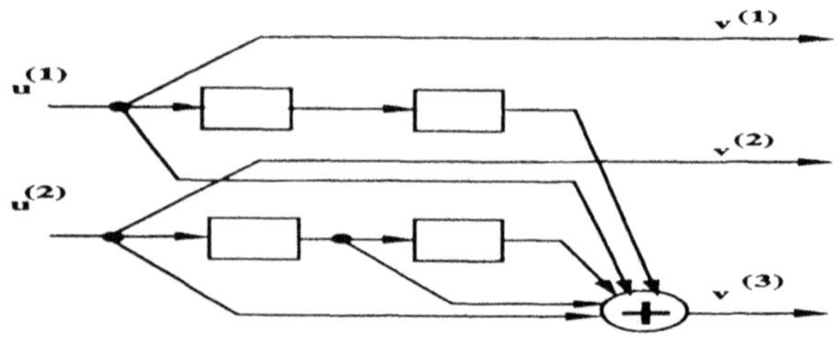


**Figure 4.1** A (3, 2, 2) systematic encoder.

**Definition 4.1** *An (n, k) convolutional encoder is called systematic if* $\mathbf{v}^{(j)} = \mathbf{u}^{(j)}$ *for* $j = 1, \ldots, k$.

We have made the first $k$ encoded sequences the same as the $k$ information sequences in this definition, but there is no loss of generality since the $n$ encoded sequences can always be reordered to achieve this. The remaining $n-k$ encoded sequences are often called the *parity-check sequences*. Encoders that are not systematic are called *nonsystematic* encoders. Systematic encoders offer the advantage that in case of no errors, the information sequences can be readily recovered, i.e., no elaborate inversion step is necessary. However, for the same rate and constraint length, the distance structure of nonsystematic encoders is superior to that of systematic encoders [92]. Costello [91] showed that every convolutional encoder is equivalent to a systematic rational encoder.

**Example 4.1** Consider the (2, 1, 3) encoder in Figure 2.2. The generator polynomials for this encoder are $\mathbf{g}^{(1)}(D) = 1$ and $\mathbf{g}^{(2)}(D) = 1 + D^2 + D^3$. It is easy to verify that $\mathbf{v}^{(1)}(D) = \mathbf{u}(D)$. For a (3, 2, $m$) systematic encoder, we must have $\mathbf{g}_1^{(1)}(D) = \mathbf{g}_2^{(2)}(D) = 1$, and $\mathbf{g}_1^{(2)}(D) = \mathbf{g}_2^{(1)}(D) = 0$. A (3, 2, 2) systematic encoder with $\mathbf{g}_1^{(3)}(D) = 1 + D^2$ and $\mathbf{g}_2^{(3)}(D) = 1 + D + D^2$ is shown in Figure 4.1.

## 4.3 DEGENERATE AND NONDEGENERATE ENCODERS

Bahl and Jelinek [24] developed a criterion for identifying ($n$, 1, $m = \nu$) encoders of a given constraint length $\nu$ that are effectively of shorter constraint lengths, and defined a class of rate-$1/n$ *nondegenerate* encoders. The search for good $(2, 1, m = \nu)$ encoders can be restricted to this class.

**Definition 4.2 (Bahl and Jelinek)** *A rate-$1/n$, constraint length $\nu$, convolutional encoder is nondegenerate if (a) there exists $j_1 \in \{1, \ldots, n\}$ such that $g_0^{(j_1)} = 1$, (b) there exists $j_2 \in \{1, \ldots, n\}$ such that $g_\nu^{(j_2)} = 1$, and (c) there exist $j_1, j_2 \in \{1, \ldots, n\}$, $j_1 \neq j_2$, such that $\mathbf{g}^{(j_1)} \neq \mathbf{g}^{(j_2)}$.*

**Example 4.2** A (2, 1, 4) encoder with generator sequences $\mathbf{g}^{(1)} = 10110$ and $\mathbf{g}^{(2)} = 01110$ is degenerate, since condition (*b*) above is not met. It is, in fact, a (2, 1, 3) encoder with generator sequences $\mathbf{g}^{(1)} = 1011$ and $\mathbf{g}^{(2)} = 0111$. Similarly, a (3, 1, 2) encoder with generator sequences $\mathbf{g}^{(1)} = 101$, $\mathbf{g}^{(2)} = 101$, and $\mathbf{g}^{(3)} = 101$ is also degenerate.

## 4.4 CATASTROPHIC AND NONCATASTROPHIC ENCODERS

In Chapter 3, we saw an example of an encoder that generates a finite weight codeword for an infinite weight information sequence. There is a nonzero probability that when this codeword is transmitted over a noisy channel, a finite number of channel errors can change all the nonzero encoded digits into zeros, thereby making the decoder generate an all-zero output at the receiver. Thus, a finite number of channel errors can cause an infinite number of decoding errors. This situation, clearly undesirable, is known as *catastrophic error propagation* and is the basis of the definition of *catastrophic* encoders.[1]

Massey and Sain [321] observed that there is a polynomial encoder equivalent to any rational encoder, which can be obtained by multiplying each row of the rational encoder by the least common multiple of the denominators of its components. As mentioned in Chapter 2, a convolutional encoder can be conveniently described from a sequential circuit point of view. A polynomial encoder $\mathbf{G}(D)$ specifies the LSC as having finite memory, i.e., as having a "feedforward" realization. It is also desirable that this LSC should possess a feedforward inverse, either instantaneous or with delay, in order to avoid catastrophic error propagation. Massey and Sain [322] showed that if for a

---

[1] The term "catastrophic" in this sense was first used by Massey in a seminar at U.C.L.A. in 1969 [155].

given LSC there exists an input sequence with infinitely many nonzero digits such that the corresponding output sequence has only finitely many nonzero digits, then the LSC has no feedforward inverse, with delay or without delay. Their main result is summarized below:

**Theorem 4.1 (Massey and Sain)** *A k-input, n-output feedforward LSC has a feedforward inverse, either with delay or without delay, if and only if*

$$gcd\left[\Delta_i(D), i = 1, 2, \ldots, \binom{n}{k}\right] = D^l, \tag{4.2}$$

*for some nonnegative integer l, where* $\Delta_i(D)$, $1 \le i \le \binom{n}{k}$, *are the determinants of* $\binom{n}{k}$ *distinct* $k \times k$ *submatrices of* $\mathbf{G}(D)$.

Massey and Sain defined an *inverse with delay* $l$ for an encoder $\mathbf{G}(D)$ as a realizable LSC with an $n \times k$ generator matrix $\tilde{\mathbf{G}}^{-1}(D)$ such that $\mathbf{G}(D)\tilde{\mathbf{G}}^{-1}(D) = D^l\mathbf{I}_k$, where $\mathbf{I}_k$ is the $k \times k$ identity matrix. Such an LSC is called a right inverse of $\mathbf{G}(D)$ if $l = 0$, or a right *pseudo-inverse* for any $l$.

Encoders without a polynomial pseudo-inverse are called *catastrophic* encoders, and the ones that are not catastrophic are called *noncatastrophic* encoders. Note that systematic encoders have a trivial polynomial inverse $\tilde{\mathbf{G}}^{-1}(D) = [\mathbf{I}_k \mid \mathbf{0}]$ with zero delay, and hence are always noncatastrophic.

For $k = 1$, i.e., for rate-$1/n$ codes, (4.2) reduces to

$$gcd\left[\mathbf{g}^{(j)}(D), j = 1,\ 2,\ \ldots,\ n\right] = D^l, \tag{4.3}$$

for some nonnegative integer $l$. From (4.3) and Definition 4.2, a rate-$1/n$ nondegenerate encoder is noncatastrophic if and only if

$$gcd\left[\mathbf{g}^{(j)}(D), j = 1,\ 2,\ \ldots,\ n\right] = 1, \tag{4.4}$$

since if the left hand side is $D^l$ and $l > 0$, then $g_0^{(j)} = 0$ for $j \in (1, 2, \ldots, n)$ and the code becomes degenerate.

**Example 4.3** For the (2, 1, 2) encoder with generator polynomials $\mathbf{g}^{(1)}(D) = 1 + D$, and $\mathbf{g}^{(2)}(D) = 1 + D^2$, we have $gcd\ [\mathbf{g}^{(1)}(D),\ \mathbf{g}^{(2)}(D)] = 1 + D$, and

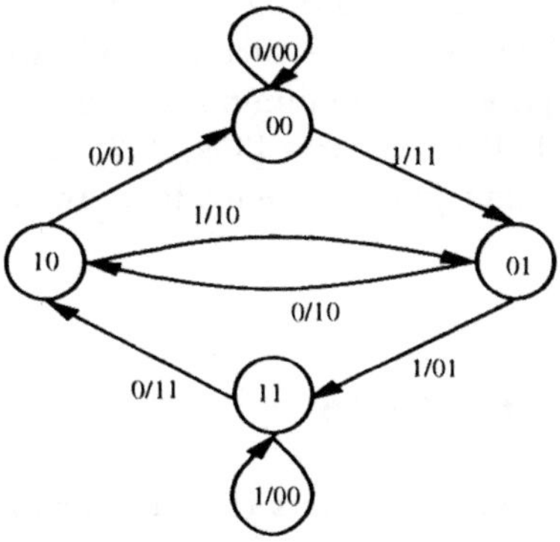


**Figure 4.2** State diagram of a (2, 1, 2) catastrophic encoder.

hence the encoder is catastrophic. Note that this encoder is nondegenerate. Let the information be $\mathbf{u}(D) = 1/(1 + D)$, i.e., the sequence $\mathbf{u} = 111\cdots$. The encoded outputs in this case are $\mathbf{v}^{(1)}(D) = \mathbf{u}(D)\,\mathbf{g}^{(1)}(D) = 1$ and $\mathbf{v}^{(2)}(D) = \mathbf{u}(D)\,\mathbf{g}^{(2)}(D) = 1 + D$. This corresponds to the composite encoded sequence $\mathbf{v} = 11\ 01\ 00\ 00\cdots$. If channel errors affect the three non-zero bits, the received sequence will be $\mathbf{r} = 00\ 00\ 00\cdots$. The decoder will produce the all-zero sequence as the estimate of the original information sequence, thereby resulting in an infinite number of decoding errors.

The state diagram of a catastrophic encoder has the following property:

**Property 4.1** *An encoder is catastrophic if and only if its state diagram has a zero-weight cycle other than the self-loop around the all-zero state.*

**Example 4.4** The state diagram of the (2, 1, 2) encoder of Example 4.3 is shown in Figure 4.2. State 11 in Figure 4.2 has a self-loop which constitutes a zero-weight cycle other than the self-loop around the all-zero state.

## 4.5 MINIMAL ENCODERS

In Chapter 2, a question was raised regarding the choice of an encoder in the equivalence class of all the encoder that generate a given code. Nondegeneracy and noncatastrophicity are two desirable properties. A further choice can be made based on the "efficiency" of the encoder realization. It is desirable to select an encoder that requires a minimum number of delay or memory elements. Forney [147, 149, 153, 154] and others [349, 337, 421, 362, 326, 225] have developed criteria for selecting just such encoders. In this section, we provide a synopsis of their results.

### 4.5.1 Basic Encoders

Forney [147] defined a class of encoders that are always noncatastrophic.

**Definition 4.3 (Forney)** *A convolutional encoder is called basic if it is polynomial and has a polynomial right inverse.*

Every rational encoder is equivalent to a basic encoder. How does one find a basic encoder equivalent to a given encoder? This question is answered by appeal to the invariant factor theorem (given in Appendix A) and is summarized in the following algorithm [225]:

1. Compute the invariant factor decomposition $\mathbf{G}(D) = \mathbf{A}(D)\Lambda(D)\mathbf{B}(D)$. The matrix $\mathbf{B}(D)$ has a polynomial inverse.
2. Let $\mathbf{G}'(D)$ be the first $k$ rows of $\mathbf{B}(D)$. Then, $\mathbf{G}'(D)$ is a basic encoder equivalent to $\mathbf{G}(D)$, since it is polynomial, and has a polynomial right inverse consisting of the first $k$ columns of $\mathbf{B}^{-1}(D)$.

McEliece and Onyszchuk [326] have given an extended invariant factor theorem that can be used for this purpose and for getting the dual encoder $\mathbf{H}(D)$. Piret [362, Sec. 2.3] has described an alternative approach for achieving the same goals.

### 4.5.2 Minimal Encoders

Among all equivalent basic encoders, one can be found whose controller canonical form[2] requires a minimal number of memory elements.

[2] A realization in the controller canonical form has been called the *obvious* realization by Forney [147].

**Definition 4.4 (Forney)** *A basic encoder is minimal if its overall constraint length $\nu$ in the obvious realization is equal to the maximum degree $\mu$ of its $k \times k$ subdeterminants.*

Let $[\mathbf{G}(D)]_h$ denote the matrix of the row-wise highest order coefficients in $\mathbf{G}(D)$, which has a nonzero entry only in the positions $(i, j)$, where $deg\ \mathbf{g}_{ij}(D) = \nu_i$. The encoder $\mathbf{G}(D)$ is minimal if $[\mathbf{G}(D)]_h$ has full rank. Forney [147] showed that every encoder is equivalent to a minimal encoder, and if the minimal encoder has overall constraint length $\mu$, then the dual code can also be generated by a minimal encoder with overall constraint length $\mu$. An algorithm to construct a minimal encoder equivalent to a given basic encoder is given below [147, 225].

1. If $[\mathbf{G}(D)]_h$ has full rank, then $\mathbf{G}(D)$ is a minimal-basic encoding matrix and we stop; else go to next step.

2. Let $[\mathbf{r}_{i_1}], [\mathbf{r}_{i_2}], \cdots, [\mathbf{r}_{i_d}]$ denote a set of rows of $[\mathbf{G}(D)]_h$ such that $\nu_{i_d} \geq \nu_{i_j}$, $1 \leq j < d$, and $[\mathbf{r}_{i_1}] + [\mathbf{r}_{i_2}] + \cdots + [\mathbf{r}_{i_d}] = \mathbf{0}$. Let $\mathbf{r}_{i_1}, \mathbf{r}_{i_2}, \cdots, \mathbf{r}_{i_d}$ denote the corresponding set of rows of $\mathbf{G}(D)$. Add

$$D^{\nu_{i_d} - \nu_{i_1}} \mathbf{r}_1 + D^{\nu_{i_d} - \nu_{i_2}} \mathbf{r}_2 + \cdots D^{\nu_{i_d} - \nu_{i_{d-1}}} \mathbf{r}_{d-1}$$

   to the $i_d$th row of $\mathbf{G}(D)$. Call the new matrix $\mathbf{G}(D)$ and go to step 1.

In a recent semitutorial work, Johannesson and Wan [225] have pointed out the existence of basic encoders that have a minimal number of abstract states, and hence, require a minimal number of memory elements over all equivalent encoders, but are not minimal according to Definition 4.4.[3] To resolve this situation, Johannesson and Wan introduced three minimality criteria: minimal-basic encoding matrix (minimal overall constraint length over equivalent basic encoding matrices), minimal encoding matrices (minimal number of abstract states over equivalent encoding matrices), and minimal encoders (realization of a minimal encoding matrix with a minimal number of memory elements over all realizations).[4] They have shown that every minimal-basic encoding matrix is a minimal encoding matrix, and that every minimal encoding matrix is noncatastrophic. Finally, they have also shown that every systematic encoding matrix is minimal.

[3] The abstract state of an encoder $\mathbf{G}(D)$ was defined in Section 2.3.1.

[4] Recall from Chapter 2 that Johannesson and Wan have defined an encoding matrix as a realizable and delayfree generator matrix, and an encoder as a realization using an LSC of a transducer. Their definition of a minimal-basic encoding matrix is equivalent to Definition 4.4.

## 4.6 LOCALLY INVERTIBLE ENCODERS

An encoder, in order to be useful at all, must have a one-to-one correspondence between the sets of information and encoded sequences. Otherwise, a decoder would not be able to decode a received sequence even in the most desirable situation of an error-free transmission. Such an encoder is *globally invertible.* In this section, a class of encoders that are also invertible in a *local* sense is described. Encoders in this class were proposed by Bitzer, Vouk, and Dholakia, [54, 109, 111, 110, 53] and are called *locally invertible* encoders.

### 4.6.1 (2, 1, m) Locally Invertible Encoders

Consider production of encoded bits using a (2, 1, $m$) encoder. Initially, $m+1$ information bits produce two encoded bits. The next information bit results in two more encoded bits, etc. After the $j$th iteration, $(m+j)$ information bits generate $2j$ encoded bits. We are interested in establishing a one-to-one correspondence between equal length subsequences. Hence, by setting $m+j = 2j$, we get $j = m$. That is, after the $m$th iteration, we would have processed $2m$ information bits and generated $2m$ encoded bits. This encoding operation can be represented using the following definition.

**Definition 4.5** *Given the generator sequences* $\mathbf{g}^{(j)} = (g_0^{(j)}, g_1^{(j)}, \ldots, g_m^{(j)})$, $j = 1, 2$, *for a (2, 1, m) convolutional encoder, we define the reduced encoding matrix as*

$$[\mathbf{G}]_{2m\times 2m} = \begin{bmatrix} g_m^{(1)} & g_m^{(2)} & & & & & & \\ g_{m-1}^{(1)} & g_{m-1}^{(2)} & g_m^{(1)} & g_m^{(2)} & & & & \\ \vdots & \vdots & \vdots & \vdots & \ddots & & & \\ g_1^{(1)} & g_1^{(2)} & g_2^{(1)} & g_2^{(2)} & \cdots & g_m^{(1)} & g_m^{(2)} \\ g_0^{(1)} & g_0^{(2)} & g_1^{(1)} & g_1^{(2)} & \cdots & g_{m-1}^{(1)} & g_{m-1}^{(2)} \\ & & g_0^{(1)} & g_0^{(2)} & \cdots & g_{m-2}^{(1)} & g_{m-2}^{(2)} \\ & & & & \ddots & \vdots & \vdots \\ & & & & & g_0^{(1)} & g_0^{(2)} \end{bmatrix}, \qquad (4.5)$$

*where blanks denote zeros.*

Let a $2m$-bit information subsequence be denoted by $\mathbf{u}_{[i,i+2m-1]} = (u_i, u_{i+1}, \ldots, u_{i+2m-1})$, and the corresponding $2m$-bit encoded subsequence be denoted by $\mathbf{v}_{[i,i+m-1]} = (\mathbf{v}_i, \mathbf{v}_{i+1}, \ldots, \mathbf{v}_{i+m-1}) = (v_i^{(1)}, v_i^{(2)}, v_{i+1}^{(1)}, v_{i+1}^{(2)}, \ldots, v_{i+m-1}^{(1)}, v_{i+m-1}^{(2)})$, where the $i$th encoded block is $\mathbf{v}_i = (v_i^{(1)}, v_i^{(2)})$. Note that since each encoded block has two bits in it, $m$ such blocks make up the $2m$-bit encoded subsequence. A $2m$-bit information subsequence is encoded to obtain the corresponding $2m$-bit encoded subsequence as follows:

$$\mathbf{v}_{[i,i+m-1]} = \mathbf{u}_{[i,i+2m-1]}[\mathbf{G}]_{2m\times 2m}. \tag{4.6}$$

Any (2, 1, $m$) encoder can be used to encode an information sequence in this manner. An encoded sequence obtained by (4.6) can be uniquely inverted if and only if the inverse of the reduced encoding matrix $[\mathbf{G}]_{2m\times 2m}$ exists. In this case we say that the mapping between these blocks is one-to-one. Let the inverse of $[\mathbf{G}]_{2m\times 2m}$ be denoted by $[\mathbf{G}]_{2m\times 2m}^{-1}$, i.e., $[\mathbf{G}]_{2m\times 2m}[\mathbf{G}]_{2m\times 2m}^{-1} = \mathbf{I}_{2m}$, where $\mathbf{I}_{2m}$ is the $2m \times 2m$ identity matrix. A $2m$-bit information subsequence is recovered by decoding the corresponding $2m$-bit encoded subsequence as follows:

$$\hat{\mathbf{u}}_{[i,2m+i-1]} = \hat{\mathbf{v}}_{[i,i+m-1]}[\mathbf{G}]_{2m\times 2m}^{-1} \tag{4.7}$$

where $\hat{\mathbf{v}}$ and $\hat{\mathbf{u}}$ are the estimates of the encoded and the information sequences, respectively. Clearly, if $\hat{\mathbf{v}}_{[i,i+m-1]} = \mathbf{v}_{[i,i+m-1]}$, then we have $\hat{\mathbf{u}}_{[i,i+2m-1]} = \mathbf{u}_{[i,i+2m-1]}$.

**Definition 4.6** *A (2, 1, m) convolutional encoder that has an invertible reduced encoding matrix* $[\mathbf{G}]_{2m\times 2m}$ *is called a locally invertible encoder.*

The inversion operation given by (4.7) can be implemented as a fast look-up table or can be built into the decoder hardware. Note that both systematic and nonsystematic encoders may be locally invertible. The class of (2, 1, $m$) locally invertible encoders has the following property: a (2, 1, $m$) convolutional encoder is both nondegenerate and noncatastrophic if and only if it is locally invertible [109, 110]. Noncatastrophicity and nondegeneracy of (2, 1, $m$) encoders are highly desirable properties, and can be ensured if the search for good (2, 1, $m$) encoders is restricted to the class of locally invertible encoders.

**Example 4.5** The (2, 1, 3) encoder with generator sequences $\mathbf{g}^{(1)} = 1010$ and $\mathbf{g}^{(2)} = 1110$ is degenerate, but noncatastrophic, and therefore not locally invertible. The (2, 1, 2) encoder with generator sequences $\mathbf{g}^{(1)} = 101$ and $\mathbf{g}^{(2)} = 110$ is nondegenerate but catastrophic, and therefore, it is also not locally invertible.

The encoding and decoding operations given by (4.6) and (4.7), respectively, have the following interesting property:

**Property 4.2** *If the two information subsequences* $\mathbf{u}_{[i,i+2m-1]}$ *and* $\mathbf{u}_{[i+1,i+2m]}$ *that overlap in* $(2m-1)$ *bits are encoded using (4.6), then the corresponding encoded subsequences* $\mathbf{v}_{[i,i+m-1]}$ *and* $\mathbf{v}_{[i+1,i+m]}$ *overlap in* $(2m-2)$ *bits. Equivalently, if two* $2m$*-bit encoded subsequences that overlap in* $(2m-2)$ *bits are decoded using (4.7), the corresponding information subsequences overlap in* $(2m-1)$ *bits.*

**Example 4.6** Consider the (2, 1, 2) encoder with generator sequences $\mathbf{g}^{(1)} = 110$ and $\mathbf{g}^{(2)} = 111$. The reduced encoding matrix and its inverse for this encoder are, respectively,

$$[\mathbf{G}]_{4\times4} = \begin{bmatrix} 0 & 1 & 0 & 0 \\ 1 & 1 & 0 & 1 \\ 1 & 1 & 1 & 1 \\ 0 & 0 & 1 & 1 \end{bmatrix} \text{ and } [\mathbf{G}]^{-1}_{4\times4} = \begin{bmatrix} 1 & 0 & 1 & 1 \\ 1 & 0 & 0 & 0 \\ 0 & 1 & 1 & 0 \\ 0 & 1 & 1 & 1 \end{bmatrix}.$$

Let the information sequence be $\mathbf{u} = 00101101100$, in which two 0's have been prepended and appended for proper initialization and termination. Then (4.6) encodes the information sequence four bits at a time. For example, the first information subsequence 0010 produces 1111. Next, shifting in one information bit gives the second information subsequence 0101, which is encoded to get 1110, etc. Notice that the first two bits of the second encoded subsequence 1110 and the last two bits of the first encoded subsequence 1111 are the same, i.e., these subsequences overlap in $4-2=2$ bits. The entire encoded sequence is $\mathbf{v} = 111110001010001001$. The first 4-bit encoded subsequence 1111 produces 0010 as the information subsequence. Shifting in two new encoded bits gives the second encoded subsequence 1110 which inverts to the second decoded subsequence 0101. Again, notice that the first three bits of the second information subsequence 0101 and the last three bits of the first information subsequence 0010 are the same, i.e., these subsequences overlap in $4-1=3$ bits. The entire decoded sequence is $\hat{\mathbf{u}} = 00101101100$.

Let the original message $\mathbf{u}$ be $N$ bits long. Let the sequence obtained after prepending and appending $m$ zeros to this information sequence be denoted by $\mathbf{u}_{[0,N+2m-1]}$, in which $\mathbf{u}_{[0,m-1]}$ and $\mathbf{u}_{[N+m,N+2m-1]}$ are zeros used for initializing and terminating the convolutional encoding process. Note that prepending zeros to the information sequence achieves the same purpose as initializing a shift-register-based encoder to the all-zero state, while appending zeros ensures termination of the encoder in the all-zero state. The information sequence $\mathbf{u}_{[0,N+2m-1]}$ is encoded $2m$-bits at a time by using (4.6) to obtain the composite encoded sequence $\mathbf{v}_{[0,N+m-1]}$.

The encoded sequence can be separated into two subsequences $\mathbf{v}_I$ and $\mathbf{v}_{II}$ by assigning non-overlapping $2m$-bit subsequences alternatively to one and then the other. Thus, we have $\mathbf{v}_I = (\mathbf{v}_{[0,m-1]}, \mathbf{v}_{[2m,3m-1]}, \ldots, \mathbf{v}_{[n-m,n-1]})$ and $\mathbf{v}_{II} = (\mathbf{v}_{[m,2m-1]}, \mathbf{v}_{[3m,4m-1]}, \ldots, \mathbf{v}_{[n,n+m-1]})$ as the two encoded sequences. An important property of these two sequences is that any one of them can be used to recover the original information sequence $\mathbf{u}_{[m,n+m-1]}$ by using (4.7) for decoding. Consider the inversion of the sequence $\mathbf{v}_I$. The encoded subsequence $\mathbf{v}_{[0,m-1]}$ decodes to $\mathbf{u}_{[0,2m-1]}$, $\mathbf{v}_{[2m,3m-1]}$ decodes to $\mathbf{u}_{[2m,4m-1]}$, and finally, $\mathbf{v}_{[n-m,n-1]}$ decodes to $\mathbf{u}_{[n-m,n+m-1]}$. Thus, $\mathbf{v}_I$ can be inverted to get $\mathbf{u}_{[0,n+m-1]}$, from which the original message $\mathbf{u}_{[m,n+m-1]}$ can be easily obtained. The sequence $\mathbf{v}_{II}$ can also be similarly inverted.

The lengths of the two sequences $\mathbf{v}_I$ and $\mathbf{v}_{II}$ may not always be equal. An $N$-bit information sequence is encoded into a sequence of length $2(N+m)$. The number of $2m$-bit adjacent subsequences in this encoded sequence is $\frac{2(M+m)}{2m} = 1 + \frac{N}{m}$. This number would be an integer if $m$ divides $N$, and would be an even integer if $N = (2l-1)m$ for some integer $l$. When both these requirements are met, the lengths of the two encoded sequences $\mathbf{v}_I$ and $\mathbf{v}_{II}$ are equal. This property of locally invertible encoders is summarized below.

**Property 4.3** *For a rate one-half invertible convolutional code with constraint length $\nu$, let an $N$-bit information sequence $\mathbf{u}$ be encoded into the composite sequence $\mathbf{v}$. Then, $\mathbf{v}$ can be split into two sequences $\mathbf{v}_I$ and $\mathbf{v}_{II}$ such that the information sequence $\mathbf{u}$ can be completely recovered from either $\mathbf{v}_I$ or $\mathbf{v}_{II}$ by using (4.7).*

**Example 4.7** Consider the (2, 1, 2) encoder of Example 4.6. Let $\mathbf{u} = 1011001$ $1011001$ be a 14-bit information sequence. After prepending and appending zeros for initialization, the message used for encoding is $\mathbf{u}_{[0,17]}$ 00101100 1101100100, in which $\mathbf{u}_{[2,15]}$ is the original information. The corresponding composite encoded sequence is $\mathbf{v}_{[0,15]} =$ 1111 1000 1001 1100 1010 0010 0111 1101. It is possible to completely recover $\mathbf{u}_{[0,15]}$ from the encoded subsequences $\mathbf{v}_{[0,1]}$, $\mathbf{v}_{[4,5]}$, $\mathbf{v}_{[8,9]}$, and $\mathbf{v}_{[12,13]}$. Similarly, the encoded subsequences $\mathbf{v}_{[2,3]}$, $\mathbf{v}_{[6,7]}$, $\mathbf{v}_{[10,11]}$, and $\mathbf{v}_{[14,15]}$ can also be used to completely recover $\mathbf{u}_{[2,17]}$, from which the original message $\mathbf{u}_{[2,15]}$ can be easily obtained. Hence, we have $\mathbf{v}_I = 1111100110100111$ and $\mathbf{v}_{II} = 1000110000101101$.

### 4.6.2 (n, 1, m) Locally Invertible Encoders

Local invertibility implies existence of a one-to-one correspondence, in the time domain, between equal length information and encoded subsequences of some finite length $w$. This correspondence is specified by a full rank $w \times w$ matrix $[\mathbf{G}]_{w\times w}$ obtained from the semi-infinite generator matrix $\mathbf{G}$. The matrix $[\mathbf{G}]_{w\times w}$ can be used to generate the encoded sequence and its inverse $[\mathbf{G}]^{-1}_{w\times w}$ can be used to recover the information sequence. The first $m$ bits of a $w$-bit information block represent the current encoder state. The remaining $(w-m)$ bits generate $n(w-m)$ encoded bits. For a $(n,\ 1,\ m)$ encoder, the encoded bits involved in the one-to-one correspondence are chosen from this block of $n(w-m)$ encoded bits. Similar to $(2,\ 1,\ m)$ encoders, $(n,\ 1,\ m)$ encoders can be shown to have the following properties [109, 53]: $(n,\ 1,\ m)$ convolutional encoders are both noncatastrophic and nondegenerate if and only if they are locally invertible.

A problem investigated by many researchers [24, 345, 386, 149] is the zero-run problem. It can be stated as follows [149]: if an encoder $\mathbf{G}$ does not start in the zero state, for how many consecutive time units $\tau$ can its outputs $\mathbf{v}_t = (v_t^{(1)},\ \ldots,\ v_t^{(n)})$, $1 \leq t \leq \tau$, be all-zero? For $(n,\ 1,\ m)$ encoders, upper and lower bounds on this number have been provided by Odenwalder [345] and Rosenberg [386], respectively. These results were further generalized by Forney [149] to $(n,\ k,\ m)$ encoders by exploring the structure of the dual encoder. For $(n,\ 1,\ m)$ encoders, these bounds can also be derived using the property of local invertibility [53].

### 4.6.3 (n, k ≥ 2, m) Locally Invertible Encoders

In this section, $(n,\ k \geq 2,\ m)$ locally invertible encoders are defined using the single shift register notation introduced in Section 2.3.1. The results on the noncatastrophicity and nondegeneracy of $(n,\ k \geq 2,\ m)$ locally invertible encoders are still outstanding.

Let the $n$ generator sequences of an $(n,\ k \geq 2,\ m)$ encoder be $G^{(j)} = (G_0^{(j)},\ G_1^{(j)},\ \ldots,\ G_{K-1}^{(j)})$, $j = 1,\ \ldots,\ n$, where $K = \nu + k$ is the input constraint length. Consider the encoding operation. Initially, $K$ information bits produce $n$ encoded bits. The next set of $k$ information bits result in $n$ more encoded bits, etc. After the $j$th iteration, $K + (j-1)k$ information bits generate $jn$

encoded bits. To obtain a one-to-one correspondence between information and encoded subsequences, we set $K + (j-1)k = jn$, and get $j = \frac{(K-k)}{(n-k)}$. That is, after the $j$th iteration,

$$w = jn = \frac{n(K-k)}{(n-k)} \tag{4.8}$$

information bits have been used in generating $w$ encoded bits. The encoders that satisfy this property are those that result in $w$ being a positive integer. Note that for a (2, 1, $m = \nu$), we have $K = \nu + 1 = m + 1$ and $w = 2m$, as shown in Section 4.6.1. Further, if $k = n - 1$, $w = n\nu$ is always an integer. In this section, only encoders with $k = n-1$ and $\nu = lk$, for some integer $l$, will be considered. A $(n, n-1, m)$ locally invertible encoder, with $K = \nu + k = (l+1)k$, can be defined using the following reduced encoding matrix.

**Definition 4.7** *Given the generator sequences $G^{(j)}$, $j = 1, \ldots, n$, of a $(n, k = n-1, m)$ convolutional encoder, we define the reduced encoding matrix $[\mathbf{G}]_{w\times w}$ as*

$$\begin{bmatrix}
G^{(1)}_{K-1} & \cdots & G^{(n)}_{K-1} & & & & & & \\
\vdots & & \vdots & & & & & & \\
G^{(1)}_{K-1-k} & \cdots & & G^{(1)}_{K-1} & \cdots & G^{(n)}_{K-1} & & & \\
\vdots & & \vdots & \vdots & & \vdots & & & \\
G^{(1)}_{0} & \cdots & G^{(n)}_{0} & \vdots & & \vdots & \ddots & & \\
 & & & \vdots & & \vdots & & G^{(1)}_{K-1} & \cdots & G^{(n)}_{K-1} \\
 & & & G^{(1)}_{0} & \cdots & G^{(n)}_{0} & & \vdots & & \vdots \\
 & & & & & & \ddots & \vdots & & \vdots \\
 & & & & & & & G^{(1)}_{0} & \cdots & G^{(n)}_{0}
\end{bmatrix}, \tag{4.9}$$

*where blanks denote zeros.*

A $w$-bit encoded subsequence $\mathbf{v}_{[i,i+\nu-1]}$ consists of $\nu$ $n$-tuples $\mathbf{v}_t = (v_t^{(1)}, \ldots, v_t^{(n)})$. Similarly, a $w$-bit information subsequence $\mathbf{u}_{[i,i+nl-1]}$ consists of $nl$ $k$-tuples $\mathbf{u}_t = (u_t^{(1)}, \ldots, u_t^{(k)})$. The encoding operation is given by

$$\mathbf{v}_{[i,i+\nu-1]} = \mathbf{u}_{[i,i+nl-1]}[\mathbf{G}]_{w\times w}. \tag{4.10}$$

Let the inverse of $[\mathbf{G}]_{w\times w}$ be denoted by $[\mathbf{G}]^{-1}_{w\times w}$, i.e., $[\mathbf{G}]_{w\times w}[\mathbf{G}]^{-1}_{w\times w} = \mathbf{I}_w$, where $\mathbf{I}_w$ is the $w \times w$ identity matrix. A $w$-bit information subsequence is recovered by decoding the corresponding $w$-bit encoded subsequence as follows:

$$\hat{\mathbf{u}}_{[i,i+nl-1]} = \hat{\mathbf{v}}_{[i,i+\nu-1]}[\mathbf{G}]^{-1}_{w\times w}. \tag{4.11}$$

From (4.10), it follows that a sliding window of length $w$ can be used to select the current $w$-bit information subsequence, and the sliding length in each time unit is $k$ bits. Note that such an encoder would produce overlapping encoded subsequences, with the last $w-n$ bits of an encoded subsequences overlapping the first $w-n$ bits of the succeeding encoded subsequences. The overlapping bits must be the same.

**Property 4.4** *If two $w$-bit information subsequences $\mathbf{u}_{[i,i+nl-1]}$ and $\mathbf{u}_{[i+1,i+nl]}$ that overlap in $(w-k)$ bits are encoded using (4.10), the corresponding encoded subsequences $\mathbf{v}_{[i,i+\nu-1]}$ and $\mathbf{v}_{[i+1,i+\nu]}$ overlap in $(w-n)$ bits. Equivalently, if two $w$-bit encoded subsequences that overlap in $(w-n)$ bits are decoded using (4.11) the corresponding information subsequences overlap in $(w-k)$ bits.*

This property is used in Chapter 11 to develop a lost packet recovery scheme using locally invertible encoders.

# 4.7 SOME GOOD (2, 1, m) ENCODERS

## 4.7.1 Complementary Encoders

Bahl and Jelinek defined a class of (2, 1, $m$) encoders that includes encoders with good distance properties in it.

**Definition 4.8 (Bahl and Jelinek)** *A binary rate one-half convolutional encoder with constraint length $\nu$ is complementary if and only if $g_0^{(1)} = g_0^{(2)} = g_\nu^{(1)} = g_\nu^{(2)} = 1$, and $g_j^{(1)} = g_j^{(2)} + 1$, $1 \le j \le \nu - 1$.*

**Example 4.8** The $(2,1,2)$ encoder with generator sequences $\mathbf{g}^{(1)} = (101)$ and $\mathbf{g}^{(2)} = (111)$ is complementary. In fact, this is the only complementary encoder with $\nu = 2$.

A rate-$1/n$ convolutional encoder is called *antipodal* if all the generator polynomials have zero delay and maximum degree, i.e., if $g_0^{(j)} = g_\nu^{(j)}$ for $j = 1, \ldots, n$.

Thus, complementary encoders are antipodal. A property of rate one-half antipodal encoders is that the encoded outputs on outgoing branches of the state diagram are binary complements. For instance, see Figure 2.5.

Complementary encoders with long constraint lengths are easily found since their search can be restricted to a small subset of the set of all (2, 1, $m$) encoders. All complementary encoders with $\nu \geq 2$ are nondegenerate. However, encoders with larger free distance than those of complementary encoders have been found (see Section 4.12). In addition, longer complementary encoders have catastrophic-like properties [200]. Even some short constraint length complementary encoders are catastrophic. For instance, the complementary encoder with generator sequences $\mathbf{g}^{(1)} = 1001$ and $\mathbf{g}^{(2)} = 1111$ is catastrophic (see exercises).

## 4.7.2 Quick-Look-In Encoders

Quick-look-in (QLI) encoders, defined by Massey and Costello [318], are nonsystematic convolutional encoders satisfying

$$\mathbf{g}^{(1)}(D) + \mathbf{g}^{(2)}(D) = D, \tag{4.12}$$

and $g_0^{(1)} = g_0^{(2)} = g_m^{(1)} = g_m^{(2)} = 1$. Thus, the two generator sequences differ only in the second position. A QLI encoder allows a codeword to be easily inverted, using a simple modulo-2 addition, to obtain the information polynomial $\mathbf{u}(D)$, since $\mathbf{v}^{(1)}(D)+ \mathbf{v}^{(2)}(D) = \mathbf{u}(D)\mathbf{g}^{(1)}(D)+ \mathbf{u}(D)\mathbf{g}^{(2)}(D) = \mathbf{u}(D)(\mathbf{g}^{(1)}(D)+ \mathbf{g}^{(2)}(D)) = D\mathbf{u}(D)$. Hence, QLI encoders are "systematic-like" nonsystematic encoders. Also, QLI encoders are always noncatastrophic (see exercises). Helgert [192, 193] studied a further generalization of QLI encoders in which

$$\mathbf{g}^{(1)}(D) + \mathbf{g}^{(2)}(D) = D^l \tag{4.13}$$

for some nonnegative integer $l$. QLI encoders have free distances that lag behind those of the best nonsystematic encoders as the constraint length of the encoders is increased (see Section 4.12).

## 4.7.3 Easy-Look-In Encoders

Another class of encoders that can be inverted quickly are the easy-look-in (ELI) encoders proposed by Johannesson and Paaske [221, 224]. ELI encoders satisfy $(1 + D)\mathbf{g}^{(1)}(D)+ D\mathbf{g}^{(2)}(D) = 1$.

# 4.8 PUNCTURED CODES AND ENCODERS

Recall from Chapter 2 that a rate $R = k/n$ convolutional code can be generated by many rate $R = k/n$ encoders. In this section, we describe a process known as *puncturing* that allows us to generate codes of different rates by using a single encoder. A punctured code is obtained by periodically deleting (or puncturing) encoded digits from ordinary encoded sequences. Puncturing has the effect of reducing the number of encoded digits corresponding to the information digits, i.e., of increasing the code rate. Thus, a low-rate encoder can be used to generate many high-rate codes by appropriately selecting the puncturing pattern (also known as the deletion map or the perforation pattern). The low-rate encoder is called the *original* encoder, while the high-rate encoder obtained after puncturing is the *punctured* encoder. If a rate-$1/n$ original encoder is punctured by deleting some of the $nP$ encoded bits corresponding to $P$ information bits, then $P$ is called the *puncturing period*. The puncturing pattern can be represented as an $n \times P$ matrix $\mathbf{P}$ whose elements are 1's and 0's, with a 1 indicating inclusion and a 0 indicating deletion.

**Example 4.9** Let the encoded bits generated by the (2, 1, 2) original encoder with generator polynomials $1 + D^2$ and $1 + D + D^2$ be grouped in blocks of four bits. Each such block corresponds to two information bits. By deleting one bit (say, every third) from the block of four, we are left with three encoded bits corresponding to two information bits. This gives us an $R = 2/3$ punctured encoder. The puncturing matrix in this case is $\mathbf{P} = \begin{bmatrix} 1 & 0 \\ 1 & 1 \end{bmatrix}$, where a column in $\mathbf{P}$ corresponds to an encoded block. The first column $\begin{bmatrix} 1 \\ 1 \end{bmatrix}$ means both bits of the first encoded block are included, and the second column $\begin{bmatrix} 0 \\ 1 \end{bmatrix}$ indicates that the first bit of the second encoded block is deleted.

While it is possible to generate a punctured encoder of any rate, we will be concerned with punctured encoders of rate-$k/(k+1)$ in this section. A (high) rate-$k/(k+1)$, $k \geq 2$, punctured encoder is specified by a set of $(k+1)$ generator polynomials

$$\{P^{(1)}(D),\ P^{(2)}(D),\ \ldots,\ P^{(k+1)}(D)\}$$

where

$$P^{(j)}(D) = p_0^{(j)} + p_1^{(j)} D + p_2^{(j)} D^2 + \cdots + p_\nu^{(j)} D^\nu .$$

Let $b \leq k+1$ be the number of distinct generators in the set $\{P^{(j)}(D)\}$. Then, the punctured encoder can be generated by periodically deleting bits from the output of a (low) rate-$1/b$ encoder. The outputs corresponding to $P^{(1)}(D)$ and $P^{(2)}(D)$ are transmitted on the same branch of the trellis and the outputs corresponding to each of the other generators are successively transmitted on separate branches. The constraint length of this punctured encoder is the same as that of the rate-$1/b$ encoder.

**Example 4.10** Consider an $R = 2/3$ punctured encoder with constraint length $\nu = 3$ and generator polynomials $P^{(1)}(D) = P^{(3)}(D) = 1 + D^2 + D^3$ and $P^{(2)}(D) = 1+D^3$. There are only two distinct generators, $P^{(1)}(D)$ and $P^{(2)}(D)$, and they correspond to the (2, 1, 3) original encoder which is punctured to obtain the $R = 2/3$ punctured encoder.

It is possible to find a rate-$k/(k+1)$ *ordinary* (non-punctured) encoder that is equivalent to the punctured encoder, i.e., they both generates the same code. Many properties of a punctured encoder can be determined by examining those of the equivalent ordinary encoder. A technique for finding the equivalent ordinary encoder is described next [68, 207].

Let the function $\alpha(l)$ be defined by $\alpha(l) = \begin{cases} 1, & \text{for } l = 1 \\ 0, & \text{for } l \neq 1 \end{cases}$, and let the function $\beta(i,\ j)$ be defined by $\beta(i,\ j) = i+j+\alpha(j)-2$, where $l,\ i,\ j$ are integers. Recall from Chapter 2 that $[i]_k$ is the congruence class of $i$ (mod $k$), that is the set of integers of the form $i + lk$ for $l = \cdots -1, 0, 1 \cdots$, and that $P^{(j)}_{[i]_k}(D^r)$ is the polynomial obtained from $P^{(j)}(D)$ by selecting only the $[i]_k$ degree terms and then substituting $D^r$ for $D$. The following theorem is due to Hole [207].

**Theorem 4.2 (Hole)** *The generator matrix* $\mathbf{G}(D) = [\mathbf{g}_i^{(j)}(D)]$ *of a rate-*$k/(k+1)$ *punctured convolutional encoder with generator polynomials* $\{P^{(1)}(D), P^{(2)}(D), \ldots, P^{(k+1)}(D)\}$ *is given by* $\mathbf{g}_i^{(j)}(D) = D^{(k-\beta(i,j))/k}\, P^{(j)}_{[\beta(i,j)]_k}(D^{1/k})$. *The generators* $P^{(j)}(D)$ *are obtained from* $\mathbf{G}(D)$ *by* $P^{(j)}(D) = \sum_{i=1}^{k} D^{\beta(i,j)-k}$ $\mathbf{g}_i^{(j)}(D^k)$.

Thus, an equivalent ordinary encoder can always be obtained from a rate-$k/(k+1)$ punctured encoder.

**Example 4.11** Consider the rate two-thirds ($k = 2$) punctured encoder obtained by puncturing every third bit from the output of a rate one-half original encoder with $\nu = 2$ and generator polynomials $1 + D^2$ and $1 + D + D^2$. The generator polynomials of the punctured encoder are $P^{(1)}(D) = 1 + D^2$ and $P^{(2)}(D) = P^{(3)}(D) = 1 + D + D^2$. We have and $\beta(1, 1) = 1$, $\beta(1, 2) = 1$, $\beta(1, 3) = 2$, $\beta(2, 1) = 2$, $\beta(2, 2) = 2$, and $\beta(2, 3) = 3$. Hence,

$$\mathbf{g}_1^{(1)}(D) = D^{(2-\beta(1,1))/2} P^{(1)}_{[\beta(1,1)]_2}(D^{1/2}) = D^{1/2} P^{(1)}_{[1]_2}(D^{1/2}) = 0,$$

$$\mathbf{g}_1^{(2)}(D) = D^{(2-\beta(1,2))/2} P^{(2)}_{[\beta(1,2)]_2}(D^{1/2}) = D^{1/2} P^{(2)}_{[1]_2}(D^{1/2}) = D,$$

and

$$\mathbf{g}_1^{(3)}(D) = D^{(2-\beta(1,3))/2} P^{(3)}_{[\beta(1,3)]_2}(D^{1/2}) = P^{(3)}_{[2]_2}(D^{1/2}) = 1 + D.$$

Similarly,

$$\mathbf{g}_2^{(1)}(D) = D^{(2-\beta(2,1))/2} P^{(1)}_{[\beta(2,1)]_2}(D^{1/2}) = P^{(1)}_{[2]_2}(D^{1/2}) = 1 + D,$$

$$\mathbf{g}_2^{(2)}(D) = D^{(2-\beta(2,2))/2} P^{(2)}_{[\beta(2,2)]_2}(D^{1/2}) = P^{(2)}_{[2]_2}(D^{1/2}) = 1 + D,$$

and

$$\mathbf{g}_2^{(3)}(D) = D^{(2-\beta(2,3))/2} P^{(3)}_{[\beta(2,3)]_2}(D^{1/2}) = D^{-1/2} P^{(3)}_{[3]_2}(D^{1/2}) = 1.$$

That is, the matrix

$$\mathbf{G}(D) = \begin{bmatrix} 0 & D & 1+D \\ 1+D & 1+D & 1 \end{bmatrix}$$

is the polynomial generator matrix of the equivalent ordinary rate two-thirds encoder with $\nu = 2$.

**Example 4.12** Given $\mathbf{G}(D)$, we can use Theorem 4.2 to obtain $P^{(j)}(D)$. Consider

$$\mathbf{G}(D) = \begin{bmatrix} 0 & D & 1+D \\ 1+D & 1+D & 1 \end{bmatrix}.$$

We have

$$\begin{aligned} P^{(1)}(D) &= D^{\beta(1,1)-2}\mathbf{g}_1^{(1)}(D^2) + D^{\beta(2,1)-2}\mathbf{g}_2^{(1)}(D^2) \\ &= 1 + D^2, \\ P^{(2)}(D) &= D^{\beta(1,2)-2}\mathbf{g}_1^{(2)}(D^2) + D^{\beta(2,2)-2}\mathbf{g}_2^{(2)}(D^2) \\ &= D^{-1}D^2 + 1 + D^2 = 1 + D + D^2, \end{aligned}$$

and

$$\begin{aligned} P^{(3)}(D) &= D^{\beta(1,3)-2}\mathbf{g}_1^{(3)}(D^2) + D^{\beta(2,3)-2}\mathbf{g}_2^{(3)}(D^2) \\ &= 1 + D^2 + D(1) = 1 + D + D^2. \end{aligned}$$

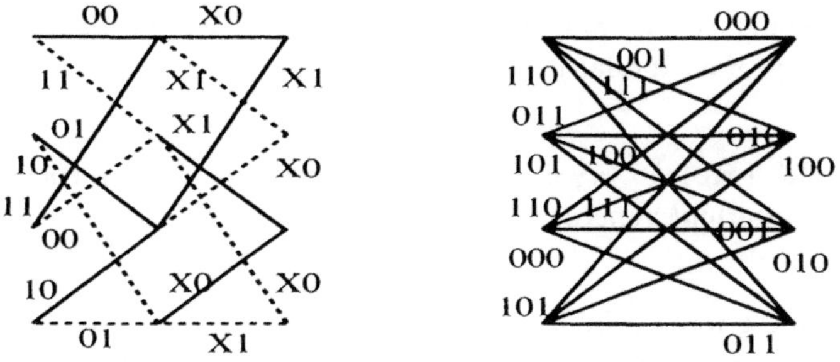


**Figure 4.3** An ordinary rate-1/2 trellis with punctured bits and the equivalent rate-2/3 trellis.

Note that these generator polynomials are the same as those in the previous example.

The puncturing process can also be described using the trellis diagram. Figure 4.3 shows two stages of the original trellis with punctured bits and a single stage of the trellis equivalent to the punctured trellis. The first output bits of the second stage are deleted, as indicated by the letter 'X'. The trellis for the equivalent ordinary encoder is obtained by merging the two stages of the original stages, with the punctured bits removed.

It is natural to ask whether the punctured encoder has any desirable properties. First, we would like to know if the punctured code is catastrophic. If the rate-$1/b$ original encoder is catastrophic, the rate-$k/(k+1)$ punctured encoder is also catastrophic. However, if the original encoder is noncatastrophic, it is still possible for the punctured encoder to be catastrophic [207]. The deletion map has to be carefully chosen to avoid this situation. Hole [206] has given an algorithm based on Property 4.1 to determine catastrophicity of a punctured encoder. The algorithm assumes that the original encoder is antipodal and searches the state diagram of the punctured encoder for non-trivial zero-weight cycles.

If both the punctured and the equivalent ordinary encoders have the same constraint length and the ordinary encoder is minimal, the punctured encoder is also minimal. When the punctured encoder is antipodal, the punctured encoder and ordinary encoder always have the same constraint length [207]. Therefore, a rate-$k/(k+1)$ antipodal punctured encoder is minimal when the corresponding ordinary encoder is minimal. Hole [208] has obtained the following result on the minimality of antipodal punctured encoders:

**Theorem 4.3 (Hole)** *Any noncatastrophic rate-$k/(k+1)$ antipodal punctured encoder is a minimal punctured encoder.*

The algorithm given by Hole [206] can be used to determine whether or not an antipodal punctured encoder is noncatastrophic, and hence, to determine whether or not such an encoder is minimal. Punctured encoders listed by many researchers [68, 273, 205] are all noncatastrophic antipodal encoders. Therefore, ordinary or punctured encoders with smaller constraint lengths that generate the same punctured codes do not exist [208].

Puncturing reduces the free distance of the code. However, if the free distance of the punctured code of a given rate and constraint length is comparable to that of an ordinary code of the same rate and constraint length, puncturing is useful since a simpler decoder structure results. Many researchers have found punctured codes with distance properties comparable to the original codes of the same rate (see Section 4.12). This situation has resulted in the use of punctured encoders in commercial products [367]. The reader may refer to [68, 487, 486, 273, 205], [177, 178, 180, 241, 341], [171, 26, 27], and [379, 329] for further details on punctured codes and encoders.

## 4.9 RATE-COMPATIBLE (RCC) ENCODERS

Two encoders of different rates are said to be *rate-compatible* if the redundant symbols generated by the higher rate encoder are also generated by the lower rate encoder. The lower rate encoder, of course, generates additional redundant symbols. A family of encoders of different rates is said to be rate-compatible if any pair of encoders in it are rate-compatible. The rate-compatibility criterion was introduced by Hagenauer [177, 178].

### *RCPC Encoders*

Hagenauer defined a class of *rate-compatible punctured convolutional* (RCPC) codes and showed them to be useful in hybrid ARQ and unequal error protection

[176, 180]. A family of RCPC encoders is described by the original $(n,\ 1,\ m)$ encoder, and $n \times P$ puncturing matrices $\mathbf{a}(l) = [a_{ij}(l)]$, with $a_{ij}(l) \in \{0,\ 1\}$ where 0 implies puncturing. Here $P$ is the puncturing period. The encoding rates given by

$$R = \frac{P}{P+l}, \quad l = 1,\ \ldots,\ (n-1)P \tag{4.14}$$

are between $P/(P+1)$ and $1/n$. The rate-compatibility restriction implies the following rule: if $a_{ij}(l_0) = 0$ then $a_{ij}(l) = 0$ for all $l \le l_0 \le (n-1)P - 1$.

**Example 4.13** The (2, 1, 2) encoder with generator sequences $\mathbf{g}^{(1)} = 101$ and $\mathbf{g}^{(2)} = 111$ can be punctured with a period $P = 4$ and the puncturing matrix

$$\mathbf{a}(1) = \begin{bmatrix} 1 & 1 & 1 & 0 \\ 1 & 0 & 0 & 1 \end{bmatrix},$$

where $l = 1$. The rate of the punctured encoder is $R = 4/5$. A family of RCPC encoders of progressively lower rates 4/6, 4/7, and 4/8 by letting $l$ =2, 3, and 4, respectively. The corresponding puncturing matrices are

$$\mathbf{a}(2) = \begin{bmatrix} 1 & 1 & 1 & 0 \\ 1 & 1 & 0 & 1 \end{bmatrix}, \mathbf{a}(3) = \begin{bmatrix} 1 & 1 & 1 & 1 \\ 1 & 1 & 0 & 1 \end{bmatrix}, \mathbf{a}(4) = \begin{bmatrix} 1 & 1 & 1 & 1 \\ 1 & 1 & 1 & 1 \end{bmatrix},$$

respectively.

## *RCRC Encoders*

A class of *rate-compatible repetition convolutional* (RCRC) encoders that are derived from a rate-$(P-1)/2(P-1)$ RCPC encoder was defined by Kallel and Haccoun [241]. A rate-1/2 encoder can be viewed as an encoder of rate-$b/2b$ if the information bits to be encoded are considered $b$ bits at a time to get $2b$ encoded bits. A repetition encoder duplicates (single or multiple duplications) a few bits from every $2b$ encoded bits. The resulting repetition encoder can be represented by a *repetition matrix* $\mathbf{Q}$, with two rows and $b$ columns, and each element indicates the number of duplications of each of the $2b$ encoded bits. The sum of the $2b$ elements of $\mathbf{Q}$ is the number of encoded bits per $b$ information bits.

Two repetition encoders obtained from the same original encoder are said to be rate-compatible if any element of the repetition matrix of the lower rate encoder is equal to or greater than its corresponding element of the repetition matrix of the higher rate encoder.

Thus, an original rate-1/2 encoder can be used to generate a high-rate punctured encoder by using a puncturing matrix, or to generate a low-rate repetition encoder by using a repetition matrix. The elements of **P** are only zeros or ones, corresponding to puncturing or not puncturing the corresponding encoded bits, while the elements of **Q** are either equal to one or greater than one, corresponding to the number of duplications of the corresponding encoded bit. The RCPC and the RCRC encoders together form the class of *rate-compatible convolutional* (RCC) encoders. The use of RCC encoders in variable-redundancy hybrid ARQ is described in Chapter 10.

## 4.10 UNIFORM CODES

Consider a $(q^m, 1, m)$ systematic encoder with generator sequences $\mathbf{g}^{(j)}$, $j = 1, \ldots, q^m$, where $g_0^{(j)} = 1$ and $\mathbf{g}_{[1,m]}^{(j)}$ are all distinct $q$-ary $m$ tuples. The code generated by such an encoder has the property that each encoded sequence is at the same distance from every sequence that disagrees in the first information block ($u_0 \neq 0$). This code, defined by Massey [308, 309], is called a *uniform code*.

**Example 4.14** Consider the case when $q = 2$ and $m = 1$. This corresponds to a rate one-half encoder with generators $\mathbf{g}^{(1)} = 10$ and $\mathbf{g}^{(2)} = 11$. All the code subsequences of length $n_A = 2(1+1) = 4$ with $u_0 \neq 0$ are 1101 and 1110. Each of them has Hamming weight equal to 3. Hence, $d_{min} = 3$.

Let $W$ be the Hamming weight of some code sub-sequence of $n_A$ digits with $u_0 \neq 0$. The average value of $W$ is given by $W_{avg} = \frac{q-1}{q} n_A + q^{m-1}$. The output constraint length $n_A = (m+1)q^m$ for this encoder. Hence, $W_{avg} = [(m+1)(q-1)+1]q^{m-1}$. Massey [309] has shown that every code subsequence of length $n_A$ with $u_0 \neq 0$ has Hamming weight $W = W_{avg}$, and hence $d_{min} = W_{avg}$.

**Definition 4.9 (Massey)** *A convolutional code is a uniform code if $d_{min} = W_{avg}$, where $W_{avg}$ is the average Hamming weight of all code subsequences of length $n_A$ with $u_0 \neq 0$.*

Massey [309] has shown that uniform codes exist for any integer $m$ and any prime-power $q$.

**Theorem 4.4 (Massey)** *For any integer $m$ and any prime-power $q$, there exists a uniform convolutional code having $R = q^{-m}$, $n_A = (m+1)q^m$, and $d_{min} = W_{avg} = [(m+1)(q-1)+1]q^{m-1}l$*

Uniform codes can be decoded using majority-logic decoding described in Chapter 7. Other properties of uniform code are described in [434, 432]. Punctured uniform codes are discussed in [377].

## 4.11 UNIT-MEMORY ENCODERS

Lee [269] defined a class of unit-memory convolutional encoders that have the largest free distance among all encoders of the same rate and encoder state complexity. Consider a rate-$k_0/n_0$ encoder with memory order $m$, and its encoding equations at time $t$ given by

$$\mathbf{v}_t = \mathbf{u}_t\mathbf{G}_0 + \mathbf{u}_{t-1}\mathbf{G}_1 + \cdots + \mathbf{u}_{t-m}\mathbf{G}_m.$$

Such an $(n_0,\ k_0,\ m)$ encoder is equivalent to an $(n' = mn_0,\ k' = mk_0,\ m' = 1)$ encoder, with

$$\mathbf{G}'_0 = \begin{bmatrix} \mathbf{G}_0 & \mathbf{G}_1 & \cdots & \mathbf{G}_{m-1} \\ 0 & \mathbf{G}_0 & \cdots & \mathbf{G}_{m-2} \\ \vdots & \vdots & & \vdots \\ 0 & 0 & \cdots & \mathbf{G}_0 \end{bmatrix} \text{ and } \mathbf{G}'_1 = \begin{bmatrix} \mathbf{G}_m & 0 & \cdots & 0 \\ \mathbf{G}_{m-1} & \mathbf{G}_m & \cdots & 0 \\ \vdots & \vdots & & \vdots \\ \mathbf{G}_1 & \mathbf{G}_2 & \cdots & \mathbf{G}_m \end{bmatrix}. \tag{4.15}$$

Encoders with memory order equal to one are called *unit-memory* encoders. Lee has defined the *state complexity* of a binary $(n_0,\ k_0,\ m)$ encoder to be $2^{mk_0}$, because there are $2^{mk_0}$ distinct states of the encoder. It is clear that the state complexity of the two equivalent encoders given above is the same. Hence, the maximum value of $d_{free}$ is achieved within the subset of convolutional encoders with $m = 1$. For other work on unit-memory encoders, see [270, 439, 230, 228].

Lauer [263] defined the state complexity $\mu$ of a unit-memory encoder to be the rank of $\mathbf{G}'_1$, which may be different from $mk_0$. Encoders with $m' = 1$ and $\mu < k'$ are called *partial-unit-memory* (PUM) encoders.

Let the constraint length $\nu(C)$ of a convolutional code $C$ be the minimum constraint length among all its encoders. Abdel-Ghaffar [4] has defined *unit constraint-length convolutional* (UCLC) code as the code with $\nu(C) = 1$. Forney [151] has defined the complexity of a maximum-likelihood decoder for a

$q$-ary rate-$k/n$ convolutional encoder with constraint length $\nu$ to be $q^{k\nu}$. UCLC codes are generated by an encoder with $\nu = 1$. Hence, they are the simplest class of convolutional codes with nonzero constraint length.

## 4.12 EXERCISES

1. Is the (2, 1, 4) encoder with $\mathbf{g}^{(1)} = 10110$ and $\mathbf{g}^{(2)} = 11110$ complementary? Is it nondegenerate?

2. Check if the (2, 1, 5) encoder with $\mathbf{g}^{(1)} = 101101$ and $\mathbf{g}^{(2)} = 110011$ is complementary. Is it catastrophic? Is it locally invertible?

3. Is the (2, 1, 4) encoder with $\mathbf{g}^{(1)} = 10011$ and $\mathbf{g}^{(2)} = 11101$ nondegenerate? Is it noncatastrophic? Is it locally invertible?

4. Determine the coding rate of a binary uniform encoder with $m = 2$. List all its generator sequences. What is its minimum distance?

5. Is the (2, 1, 2) encoder with $\mathbf{g}^{(1)} = 100$ and $\mathbf{g}^{(2)} = 001$ nondegenerate? Is it noncatastrophic? Is it locally invertible?

6. Consider the $6 \times 6$ reduced encoding matrix

$$[\mathbf{G}]_6 = \begin{bmatrix} 1 & 0 & 0 & 0 & 0 & 0 \\ 0 & 0 & 1 & 0 & 0 & 0 \\ 0 & 0 & 0 & 0 & 1 & 0 \\ 0 & 1 & 0 & 0 & 0 & 0 \\ 0 & 0 & 0 & 1 & 0 & 0 \\ 0 & 0 & 0 & 0 & 0 & 1 \end{bmatrix}.$$

   Is this matrix invertible? If so, what are the two generator sequences $\mathbf{g}^{(1)}$ and $\mathbf{g}^{(2)}$ of the corresponding (2, 1, $m$) locally invertible encoder?

7. Prove that a QLI encoder is always noncatastrophic.

8. Write a computer program to implement encoding and local inversion operations for the (2, 1, 5) encoder with $\mathbf{g}^{(1)} = 10011$ and $\mathbf{g}^{(2)} = 11101$. The one-to-one mapping length of this encoder is $w = 8$. Implement the encoding and local inversion using (4.6) and (4.7).

# APPENDIX A

## Listings of Convolutional Encoders

Many researchers have published lists of convolutional encoders that belong to many of the classes described in this chapter. In this appendix, tables of encoders of commonly used rates, lengths, and types are included. Size limitations preclude an exhaustive listing. However, references in which encoders of various rates, lengths, and types have been published are tabulated.

In this appendix, the following short forms are used: $d_f = d_{free}$, $a_f = n(d_{free})$, $c_f = n_b(d_{free})$; $d_m = d_{min}$, $a_m = n(d_{min})$; CMP = complementary, C = catastrophic, NC = Noncatastrophic.

Tables A.1 and A.2 list (2, 1, $m$) systematic encoders and are adapted from [70]. Table A.3 lists (2, 1, $m$) nonsystematic encoders and is adapted from [70]. The generator sequences are listed in left-justified octal notation. For instance, the generator sequences $\mathbf{g}^{(1)} = 1011$ and $\mathbf{g}^{(2)} = 1111$ of a (2, 1, 3) encoder are listed as 54 and 74.

Table A.4 lists rate one-third nonsystematic encoders and is adapted from [261, 89]. Table A.5 lists rate one-fourth nonsystematic encoders and is adapted from [261, 89]. Table A.6 lists rate two-thirds nonsystematic encoders and is adapted from [354]. Table A.7 lists rate three-fourths nonsystematic encoders and is adapted from [354].

Table A.8 lists rate-$k/k+1$, $k = 2, \ldots, 7$ punctured encoders and is adapted from [171].

Table A.9 lists the references in which tables of rate-$k/n$ systematic encoders of various lengths and classes can be found. Table A.10 lists the references in which tables of rate-$k/n$ nonsystematic encoders of various lengths and classes can be found. Table A.11 lists the references in which tables of encoders of some specific classes can be found. Table A.12 lists the references in which tables of distance spectra and decision depths of encoders can be found.

| $m$ | $\mathbf{g}^{(2)}$ | $d_f$ | $a_f$ | $c_f$ |
|---|---|---|---|---|
| 1 | 6 | 3 | 1 | 1 |
| 2 | 7 | 4 | 2 | 3 |
| 3 | 64 | 4 | 1 | 1 |
| 4 | 72 | 5 | 2 | 4 |
| 5 | 73 | 6 | 3 | 6 |
| 6 | 654 | 6 | 2 | 4 |
| 7 | 654 | 6 | 2 | 4 |
| 8 | 715 | 7 | 2 | 4 |
| 9 | 6714 | 8 | 5 | 13 |
| 10 | 7152 | 8 | 3 | 8 |
| 11 | 7155 | 9 | 5 | 13 |
| 12 | 67114 | 9 | 1 | 1 |
| 13 | 67116 | 10 | 5 | 13 |
| 14 | 67115 | 10 | 4 | 10 |
| 15 | 671144 | 10 | 3 | 5 |
| 16 | 671166 | 12 | 13 | 44 |
| 17 | 714447 | 11 | 4 | 12 |
| 18 | 7144474 | 12 | 5 | 14 |
| 19 | 7144616 | 12 | 3 | 7 |
| 20 | 6711455 | 12 | 2 | 6 |
| 21 | 67115144 | 12 | 2 | 8 |
| 22 | 67114552 | 14 | 12 | 47 |
| 23 | 71446165 | 14 | 10 | 41 |
| 24 | 671145434 | 15 | 7 | 27 |
| 25 | 671145452 | 15 | 6 | 22 |
| 26 | 7144761242 | 16 | 16 | 66 |
| 27 | 6711455364 | 16 | 10 | 44 |
| 28 | 7144761242 | 16 | 11 | 44 |
| 29 | 7144760535 | 18 | 22 | 89 |
| 30 | 67114545644 | 16 | 3 | 9 |
| 31 | 67114543066 | 18 | 11 | 50 |

**Table A.1** Free distance $d_f$ and spectral components $a_f$, $c_f$ of (2, 1, $m$) systematic ODP convolutional encoders.

| $m$ | $\mathbf{g}^{(2)}$ | $d_m, a_m$ | $m$ | $\mathbf{g}^{(2)}$ | $d_m, a_m$ |
|---|---|---|---|---|---|
| 1 | 6 | 3, 1 | 33 | 671145457554 | 13, 1 |
| 2 | 6 | 3, 1 | 34 | 671145457556 | 14, 34 |
| 3 | 64 | 4, 3 | 35 | 671145454470 | 14, 14 |
| 4 | 64 | 4, 1 | 36 | 6711454544704 | 14, 5 |
| 5 | 65 | 5, 5 | 37 | 6711454544676 | 14, 2 |
| 6 | 650 | 5, 2 | 38 | 6711454575564 | 15, 31 |
| 7 | 670 | 6, 11 | 39 | 67114545755644 | 15, 12 |
| 8 | 670 | 6, 5 | 40 | 67114545755712 | 15, 3 |
| 9 | 6710 | 6, 1 | 41 | 67114545755713 | 15, 1 |
| 10 | 6710 | 7, 12 | 42 | 671145457556464 | 16, 31 |
| 11 | 6711 | 7, 5 | 43 | 671145457556464 | 16, 14 |
| 12 | 67114 | 8, 29 | 44 | 671145457556153 | 16, 5 |
| 13 | 67114 | 8, 12 | 45 | 6711454575561314 | 16, 1 |
| 14 | 67115 | 8, 6 | 46 | 6711454575564666 | 17, 39 |
| 15 | 671150 | 8, 1 | 47 | 6711454575564667 | 17, 13 |
| 16 | 671144 | 9, 18 | 48 | 67114545755646674 | 17, |
| 17 | 671151 | 9, 7 | 49 | 67114545755646676 | 17, 1 |
| 18 | 6711514 | 9, 3 | 50 | 67114545755646676 | 18, 38 |
| 19 | 6711454 | 10, 31 | 51 | 671145457556466760 | 18, 16 |
| 20 | 6711454 | 10, 13 | 52 | 671145457556466760 | 18, 7 |
| 21 | 67114544 | 10, 4 | 53 | 671145457550027077 | 18, 2 |
| 22 | 67115142 | 10, 1 | 54 | 6711454575571301174 | 19, 43 |
| 23 | 67114543 | 11, 27 | 55 | 6711454575571301176 | 19, 20 |
| 24 | 671145430 | 11, 11 | 56 | 6711454575571301176 | 19, 7 |
| 25 | 671151572 | 11, 5 | 57 | 67114545755713011760 | 19, 2 |
| 26 | 671151505 | 11, 1 | 58 | 67114545755713011760 | 20, 60 |
| 27 | 6711454574 | 12, 21 | 59 | 67114545755646670367 | 20, 25 |
| 28 | 6711454306 | 12, 8 | 60 | 671145457556466703670 | 20, 10 |
| 29 | 6711454311 | 12, 2 | 61 | 671145457557130117610 | 20, 2 |
| 30 | 67114545754 | 13, 43 | 62 | 671145457557130117611 | 20, 1 |
| 31 | 67114545754 | 13, 15 | 63 | 6711454575571301176114 | 21, 25 |
| 32 | 67114545755 | 13, 4 | 64 | 6711454575571301176114 | 21, 10 |

**Table A.2** Minimum distance $d_m$ and spectral component $a_m$ for (2, 1, $m$) systematic ODP convolutional encoders ($m = 1$ - 64).

| $m$ | $\mathbf{g}^{(2)}$ | $d_m$, $a_m$ |
|---|---|---|
| 65 | 6711454575571301176114 | 21, 2 |
| 66 | 67114545755713011761144 | 22, 71 |
| 67 | 67114545755646670367016 | 22, 29 |
| 68 | 67114545755646670367017 | 22, 9 |
| 69 | 671145457556466703670170 | 22, 4 |
| 70 | 671145457556466703670170 | 22, 1 |
| 71 | 671145457557130117611463 | 23, 46 |
| 72 | 6711454575564667036701444 | 23, 16 |
| 73 | 6711454575564667036701446 | 23, 5 |
| 74 | 6711454575564667036701447 | 23, 2 |
| 75 | 67114545755713011761146370 | 24, 56 |
| 76 | 67114545755713011761146342 | 24, 20 |
| 77 | 67114545755713011761146373 | 24, 8 |
| 78 | 671145457557130117611463424 | 24, 3 |
| 79 | 671145457557130117611463432 | 25, 74 |
| 80 | 671145457557130117611463433 | 25, 33 |
| 81 | 6711454575571301176114634334 | 25, 16 |
| 82 | 6711454575564667036701447272 | 25, 4 |
| 83 | 6711454575564667036701447277 | 25, 1 |
| 84 | 67114545755646670367014472730 | 26, 41 |
| 85 | 67114545755713011761146343362 | 26, 20 |
| 86 | 67114545755713011761146343363 | 26, 6 |
| 87 | 671145457557130117611463433634 | 26, 2 |
| 88 | 671145457556466703670144727304 | 27, 62 |
| 89 | 671145457556466703670144727305 | 27, 28 |
| 90 | 6711454575564667036701447273054 | 27, 11 |
| 91 | 6711454575564667036701447273056 | 27, 5 |
| 92 | 6711454575564667036701447273357 | 27, 1 |
| 93 | 67114545755646670367014472730510 | 28, 42 |
| 94 | 67114545755646670367014472730512 | 28, 20 |
| 95 | 67114545755646670367014472730511 | 28, 5 |
| 96 | 671145457556466703670144727305110 | 28, 1 |

**Table A.2** *(continued)* $m = 65$ - $96$.

| $m$ | $\mathbf{g}^{(1)}$ | $\mathbf{g}^{(2)}$ | $d_f$ | $a_f$ | $c_f$ | Remarks |
|---|---|---|---|---|---|---|
| 2 | 7 | 5 | 5 | 1 | 1 | OMD, ODP, OFD, CMP, QLI |
| 3 | 74 | 54 | 6 | 1 | 2 | OMD, ODP, OFD, QLI |
| | 64 | | 6 | 2 | 4 | CMP |
| 4 | 62 | 56 | 7 | 2 | 4 | OMD, ODP, OFD |
| | 72 | | 7 | 2 | 4 | CMP |
| | 46 | | 7 | 2 | 4 | QLI (Non ODP) |
| | 76 | | 6 | 1 | 2 | QLI (with ODP) |
| 5 | 75 | 55 | 8 | 2 | 6 | OMD, ODP, OFD, QLI |
| | 73 | | 8 | 2 | 5 | CMP |
| 6 | 634 | 564 | 10 | 12 | 46 | OMD, ODP, OFD |
| | 754 | | 9 | 2 | 4 | CMP |
| | 454 | | 9 | 4 | 14 | QLI (Non ODP) |
| | 714 | | 8 | 1 | 4 | QLI (with ODP) |
| 7 | 626 | 572 | 10 | 1 | 6 | OMD, ODP, OFD |
| | 712 | | 10 | 3 | 10 | CMP |
| | 542 | | 9 | 1 | 1 | QLI (Non ODP) |
| | 656 | 756 | 9 | 1 | 3 | QLI (Helgert) |
| | 742 | | 9 | 1 | 1 | QLI (with ODP) |
| 8 | 751 | 557 | 12 | 10 | 40 | OMD, ODP, OFD |
| | 745 | | 11 | 2 | 4 | CMP |
| | 551 | | 10 | 1 | 2 | QLI (Non ODP) |
| | 623 | 723 | 10 | 1 | 4 | QLI (Helgert) |
| | 743 | | 9 | 1 | 5 | QLI (with ODP) |
| 9 | 7664 | 5714 | 12 | 1 | 2 | OMD, ODP, OFD |
| | 6654 | | 12 | 4 | 10 | CMP |
| | 5664 | | 11 | 3 | 11 | QLI (Non ODP) |
| | 6134 | 6334 | 11 | 2 | 8 | QLI (Helgert) |
| | 7434 | | 10 | 2 | 6 | QLI (with ODP) |
| 10 | 7512 | 5562 | 14 | 19 | 82 | OMD, ODP, OFD |
| | 7026 | | 13 | 2 | 8 | CMP |
| | 5506 | | 12 | 3 | 18 | QLI (Non ODP) |
| | 6626 | 7626 | 12 | 2 | 10 | QLI (Helgert) |
| | 7422 | | 11 | 2 | 6 | QLI (with ODP) |

**Table A.3** Free distance $d_f$ and spectral components $a_f$, $c_f$ of (2, 1, $m$) nonsystematic convolutional encoders ($m$ = 2 - 10).

| $m$ | $\mathbf{g}^{(1)}$ | $\mathbf{g}^{(2)}$ | $d_f$ | $a_f$ | $c_f$ | Remarks |
|---|---|---|---|---|---|---|
| 11 | 6643 | 5715 | 14 | 1 | 4 | OMD, ODP |
| | 7173 | 5261 | 15 | 14 | 66 | OFD |
| | 7363 | | 14 | 5 | 17 | CMP |
| | 5503 | | 13 | 8 | 44 | QLI (Non ODP) |
| | 6713 | 7713 | 13 | 6 | 36 | QLI (Helgert) |
| | 7435 | | 12 | 5 | 26 | QLI (with ODP) |
| 12 | 63374 | 47244 | 15 | 2 | 6 | OMD, ODP, OFD |
| | 53734 | 72304 | 16 | 14 | 60 | OFD |
| | 73214 | | 15 | 3 | 11 | CMP |
| | 56414 | | 14 | 10 | 64 | QLI (Non ODP) |
| | 74044 | | 11 | 2 | 1 | QLI (with ODP) |
| 13 | 45332 | 77136 | 16 | 5 | 26 | OMD, ODP, OFD |
| | 67432 | | 16 | 7 | 28 | CMP |
| | 46716 | | 14 | 3 | 20 | QLI (Non ODP) |
| | 66206 | 76206 | 14 | 2 | 6 | QLI (Helgert) |
| | 74046 | | 13 | 2 | 6 | QLI (with ODP) |
| 14 | 65231 | 43677 | 17 | 3 | 17 | OMD, ODP, OFD |
| | 63057 | 44735 | 18 | 26 | 133 | OFD |
| | 66475 | | 17 | 5 | 25 | CMP |
| | 51503 | | 15 | 6 | 26 | QLI (Non ODP) |
| | 71073 | 71473 | 15 | 3 | 29 | QLI (Helgert) |
| | 74047 | | 14 | 2 | 6 | QLI (with ODP) |
| 15 | 517604 | 664134 | 18 | 10 | 49 | OMD, ODP, OFD |
| | 533514 | 653444 | 19 | 30 | 174 | OFD |
| | 673514 | | 18 | 12 | 59 | CMP |
| | 510474 | | 16 | 11 | 74 | QLI (Non ODP) |
| | 602364 | 606364 | 16 | 6 | 32 | QLI (Helgert) |
| | 740464 | | 14 | 2 | 12 | QLI (with ODP) |
| 16 | 717066 | 522702 | 19 | 9 | 55 | OMD, ODP, OFD |
| | 626656 | 463642 | 20 | 43 | 255 | OFD |
| | 644246 | | 18 | 2 | 8 | CMP |
| | 522416 | | 16 | 2 | 10 | QLI (Non ODP) |
| | 651426 | 653426 | 16 | 1 | 8 | QLI (Helgert) |
| | 740462 | | 15 | 3 | 13 | QLI (with ODP) |

**Table A.3** *(continued)* $m$ = 11 - 16.

| $m$ | $\mathbf{g}^{(1)}$ | $\mathbf{g}^{(2)}$ | $d_f$ | $a_f$ | $c_f$ | Remarks |
|---|---|---|---|---|---|---|
| 17 | 506477 | 673711 | 20 | 12 | 90 | OMD, ODP, OFD |
| | 735505 | | 20 | 23 | 143 | CMP |
| | 454623 | | 17 | 5 | 29 | QLI (Non ODP) |
| | 601067 | 621067 | 17 | 3 | 17 | QLI (Helgert) |
| | 740463 | | 16 | 2 | 14 | QLI (with ODP) |
| 18 | 5653644 | 7746714 | 21 | 13 | 73 | OMD, ODP, OFD |
| | 4551474 | 6354344 | 22 | 65 | 418 | OFD |
| | 6756224 | | 20 | 3 | 20 | CMP |
| | 5522214 | | 18 | 6 | 44 | QLI (Non ODP) |
| | 6504664 | 6524664 | 18 | 5 | 46 | QLI (Helgert) |
| | 7404634 | | 16 | 1 | 8 | QLI (with ODP) |
| 19 | 5122642 | 7315626 | 22 | 26 | 130 | OMD, ODP, OFD |
| | 6763426 | | 20 | 1 | 8 | CMP |
| | 4517006 | | 18 | 2 | 12 | QLI (Non ODP) |
| | 6047646 | 6447646 | 19 | 11 | 83 | QLI (Helgert) |
| | 7404242 | | 15 | 1 | 1 | QLI (with ODP) |
| 20 | 6567413 | 5322305 | 22 | 2 | 14 | OMD, ODP, OFD |
| | 6243735 | | 22 | 9 | 56 | CMP |
| | 5036543 | | 19 | 4 | 28 | QLI (Non ODP) |
| | 6144363 | 6344363 | 19 | 2 | 12 | QLI (Helgert) |
| | 7404155 | | 18 | 2 | 12 | QLI (with ODP) |
| 21 | 67520654 | 50371444 | 24 | 40 | 260 | OMD, ODP, OFD |
| | 66452114 | | 22 | 3 | 24 | CMP |
| | 47653514 | | 20 | 7 | 54 | QLI (Non ODP) |
| | 60405634 | 60505634 | 20 | 4 | 40 | QLI (Helgert) |
| | 74041544 | | 18 | 2 | 18 | QLI (with ODP) |
| 22 | 67132702 | 50516146 | 24 | 25 | 190 | OMD, ODP, OFD |
| | 66237046 | | 24 | 23 | 172 | CMP |
| | 51102726 | | 20 | 1 | 6 | QLI (Non ODP) |
| | 61310166 | 61710166 | 21 | 12 | 96 | QLI (Helgert) |
| | 74042436 | | 19 | 2 | 12 | QLI (with ODP) |
| 23 | 55076157 | 75501351 | 26 | 65 | 498 | OMD, ODP, OFD |
| | 66557413 | | 24 | 7 | 42 | CMP |
| | 53171663 | | 21 | 3 | 21 | QLI (Non ODP) |
| | 70311243 | 70351243 | 21 | 2 | 10 | QLI (Helgert) |
| | 74041567 | | 19 | 1 | 11 | QLI (with ODP) |

**Table A.3** *(continued)* $m = 17$ - 23.

| $m$ | $\mathbf{g}^{(1)}$ | $\mathbf{g}^{(2)}$ | $d_f$ | $a_f$ | $c_f$ | Remarks |
|---|---|---|---|---|---|---|
| 24 | 673275374 | 506302644 | 26 | 26 | 217 | OMD, ODP, OFD |
| | 661563014 | | 26 | 49 | 414 | CMP |
| | 510676714 | | 22 | 7 | 72 | QLI (Non ODP) |
| | 603757254 | 623757254 | 22 | 3 | 20 | QLI (Helgert) |
| | 740415664 | | 20 | 1 | 6 | QLI (with ODP) |
| 25 | 665041116 | 516260772 | 27 | 24 | 214 | OMD, ODP, OFD |
| | 620113072 | | 26 | 15 | 122 | CMP |
| | 740424366 | | 20 | 1 | 6 | QLI (with ODP) |
| 26 | 670735375 | | 26 | 3 | 22 | CMP |
| | 740424175 | | 22 | 8 | 58 | QLI (with ODP) |
| 27 | 6243245414 | | 28 | 38 | 330 | CMP |
| | 7404155634 | | 22 | 2 | 14 | QLI (with ODP) |
| 28 | 7133254172 | | 28 | 13 | 92 | CMP |
| | 7404241726 | | 23 | 2 | 14 | QLI (with ODP) |
| 29 | 6166044143 | | 28 | 2 | 14 | CMP |
| | 7404154035 | | 24 | 6 | 64 | QLI (with ODP) |
| 30 | 67526462014 | | 30 | 24 | 212 | CMP |
| | 74041567514 | | 23 | 1 | 15 | QLI (with ODP) |
| 31 | 62162354046 | | 30 | 9 | 76 | CMP |
| | 74041567512 | | 25 | 5 | 51 | QLI (with ODP) |

**Table A.3** *(continued)* $m = 24$ - $31$.

| $m$ | $\mathbf{g}^{(1)}$ | $\mathbf{g}^{(2)}$ | $\mathbf{g}^{(3)}$ | $d_f$ | $a_f$ | $c_f$ |
|---|---|---|---|---|---|---|
| 2 | 5 | 7 | 7 | 8 | 2 | 3 |
| 3 | 54 | 64 | 74 | 10 | 3 | 6 |
| 4 | 52 | 66 | 76 | 12 | 5 | 12 |
| 5 | 47 | 53 | 75 | 13 | 1 | 1 |
| 6 | 554 | 624 | 764 | 15 | 3 | 11 |
| 7 | 452 | 662 | 756 | 16 | 1 | 1 |
| 8 | 557 | 663 | 711 | 18 | 5 | 11 |
| 9 | 4474 | 5724 | 7154 | 20 | 8 | 29 |
| 10 | 4726 | 5562 | 6372 | 22 | 14 | 53 |
| 11 | 4767 | 5723 | 6265 | 24 | 21 | 80 |
| 12 | 42554 | 43364 | 77304 | 24 | 10 | 27 |
| 13 | 43512 | 73542 | 76266 | 26 | 12 | 41 |

**Table A.4** Free distance $d_f$ and spectral components $a_f$, $c_f$ of (3, 1, $m$) nonsystematic convolutional encoders with maximum $d_f$.

| $m$ | $\mathbf{g}^{(1)}$ | $\mathbf{g}^{(2)}$ | $\mathbf{g}^{(3)}$ | $\mathbf{g}^{(4)}$ | $d_f$ | $a_f$ | $c_f$ |
|---|---|---|---|---|---|---|---|
| 2 | 5 | 7 | 7 | 7 | 10 | 1 | 2 |
| 3 | 54 | 64 | 64 | 74 | 13 | 2 | 4 |
| 4 | 52 | 56 | 66 | 76 | 16 | 4 | 8 |
| 5 | 53 | 67 | 71 | 75 | 18 | 3 | 6 |
| 6 | 564 | 564 | 634 | 714 | 20 | 10 | 37 |
| 7 | 472 | 572 | 626 | 736 | 22 | 1 | 2 |
| 8 | 463 | 535 | 733 | 745 | 24 | 2 | 4 |
| 9 | 4474 | 5724 | 7154 | 7254 | 27 | 4 | 10 |
| 10 | 4656 | 4726 | 5562 | 6372 | 29 | 5 | 13 |
| 11 | 4767 | 5723 | 6265 | 7455 | 32 | 14 | 49 |
| 12 | 44624 | 52374 | 66754 | 73534 | 33 | 5 | 19 |
| 13 | 42226 | 46372 | 73256 | 73276 | 36 | 19 | 74 |

**Table A.5** Free distance $d_f$ and spectral components $a_f$, $c_f$ of (4, 1, $m$) nonsystematic convolutional encoders with maximum $d_f$.

| $\nu$ | $\mathbf{g}_1^{(1)}$ | $\mathbf{g}_2^{(1)}$ | $\mathbf{g}_1^{(2)}$ | $\mathbf{g}_2^{(2)}$ | $\mathbf{g}_1^{(3)}$ | $\mathbf{g}_2^{(3)}$ | $d_f$ |
|---|---|---|---|---|---|---|---|
| 2 | 6 | 2 | 2 | 4 | 6 | 4 | 3 |
| 3 | 4 | 1 | 2 | 4 | 6 | 7 | 4 |
| 4 | 7 | 2 | 1 | 5 | 4 | 7 | 5 |
| 5 | 60 | 14 | 30 | 40 | 70 | 74 | 6 |
| 6 | 64 | 30 | 30 | 64 | 74 | 64 | 7 |
| 7 | 60 | 16 | 34 | 46 | 54 | 74 | 8 |
| 8 | 64 | 26 | 12 | 66 | 52 | 44 | 8 |
| 9 | 52 | 05 | 06 | 70 | 74 | 53 | 9 |
| 10 | 63 | 32 | 15 | 65 | 46 | 61 | 10 |

**Table A.6** Free distance $d_f = d_{free}$ of (3, 2, $m$) nonsystematic convolutional encoders.

| $\nu$ | $\mathbf{g}_1^{(1)}, \mathbf{g}_2^{(1)}, \mathbf{g}_3^{(1)}$ | $\mathbf{g}_1^{(2)}, \mathbf{g}_2^{(2)}, \mathbf{g}_3^{(2)}$ | $\mathbf{g}_1^{(3)}, \mathbf{g}_2^{(3)}, \mathbf{g}_3^{(3)}$ | $\mathbf{g}_1^{(4)}, \mathbf{g}_2^{(4)}, \mathbf{g}_3^{(4)}$ | $d_f$ |
|---|---|---|---|---|---|
| 3 | 4, 0, 0 | 4, 6, 2 | 4, 2, 5 | 4, 4, 5 | 4 |
| 5 | 6, 1, 0, | 2, 6, 2 | 2, 0, 5, | 6, 7, 5 | 5 |
| 6 | 6, 3, 2, | 1, 4, 3 | 0, 1, 7 | 7, 6, 4 | 6 |
| 8 | 70, 14, 04 | 30, 50, 10 | 20, 00, 74 | 40, 54, 40 | 7 |
| 9 | 40, 04, 34 | 14, 64, 00 | 34, 20, 60 | 60, 70, 64 | 8 |

**Table A.7** Free distance $d_f = d_{free}$ of (4, 3, $m$) nonsystematic convolutional encoders.

| Original Encoder | | Punctured Encoders | | | | | |
|---|---|---|---|---|---|---|---|
| | | $R = 2/3$ | | $R = 3/4$ | | $R = 4/5$ | |
| $m$ | $\mathbf{g}^{(1)}, \mathbf{g}^{(2)}$ | **P** | $d_f$<br>$a_f$, $c_f$ | **P** | $d_f$<br>$a_f$, $c_f$ | **P** | $d_f$<br>$a_f$, $c_f$ |
| 2 | 5, 7 | 10<br>11 | 3<br>1, 1 | 101<br>110 | 3<br>6, 15 | 1011<br>1100 | 2<br>1, 1 |
| 3 | 15, 17 | 11<br>10 | 4<br>3, 10 | 110<br>101 | 4<br>29, 124 | 1011<br>1100 | 3<br>5, 14 |
| 4 | 23, 35 | 11<br>10 | 4<br>1, 1 | 101<br>110 | 3<br>1, 1 | 1010<br>1101 | 3<br>3, 11 |
| 5 | 53, 75 | 10<br>11 | 6<br>19, 96 | 100<br>111 | 4<br>1, 3 | 1000<br>1111 | 4<br>7, 40 |
| 6 | 133, 171 | 11<br>10 | 6,<br>1, 3 | 110<br>101 | 5<br>8, 42 | 1111<br>1000 | 4<br>3, 12 |
| 7 | 247, 371 | 10<br>11 | 7<br>9, 47 | 110<br>101 | 6<br>36, 239 | 1010<br>1101 | 5<br>20, 168 |
| 8 | 561, 753 | 11<br>10 | 7<br>3, 11 | 111<br>100 | 6<br>10, 52 | 1101<br>1010 | 5<br>7, 31 |

| Original Encoder | | Punctured Encoders | | | | | |
|---|---|---|---|---|---|---|---|
| | | $R = 5/6$ | | $R = 6/7$ | | $R = 7/8$ | |
| $m$ | $\mathbf{g}^{(1)}, \mathbf{g}^{(2)}$ | **P** | $d_f$<br>$a_f$, $c_f$ | **P** | $d_f$<br>$a_f$, $c_f$ | **P** | $d_f$<br>$a_f$, $c_f$ |
| 2 | 5, 7 | 10111<br>11000 | 2<br>2, 2 | 101111<br>110000 | 2<br>4, 5 | 1011111<br>1100000 | 2<br>6, 8 |
| 3 | 15, 17 | 10100<br>11011 | 3<br>15, 63 | 100011<br>111100 | 2<br>1, 2 | 1000010<br>1111101 | 2<br>2, 4 |
| 4 | 23, 35 | 10111<br>11000 | 3<br>5, 20 | 101010<br>110101 | 3<br>14, 69 | 1010011<br>1101100 | 3<br>13, 49 |
| 5 | 53, 75 | 10000<br>11111 | 4<br>19, 100 | 110110<br>101001 | 3<br>5, 25 | 1011101<br>1100010 | 3<br>9, 60 |
| 6 | 133, 171 | 11010<br>10101 | 4<br>14, 92 | 111010<br>100101 | 3<br>1, 5 | 1111010<br>1000101 | 3<br>2, 9 |
| 7 | 247, 371 | 11100<br>10011 | 4<br>2, 7 | 101001<br>110110 | 4<br>11, 85 | 1010100<br>1101011 | 4<br>26, 258 |
| 8 | 561, 753 | 10110<br>11001 | 5<br>19, 168 | 110110<br>101001 | 4<br>2, 9 | 1101011<br>1010100 | 4<br>6, 70 |

**Table A.8** Free distance $d_f$ and spectral components $a_f$, $c_f$ of $R = 2/3$, 3/4, 4/5, 5/6, 6/7, and 7/8 punctured convolutional encoders.

| $k$ | $n$ | $m$ | Type | Reference |
|---|---|---|---|---|
| 1 | 2 | 1-15 | | Bussgang [64] |
| | | 20 | | Lin & Lyne [287] |
| | | 71 | | Costello [92] |
| | | 1-14 | ODP, OMD | Johannesson [221] |
| | | 15-35 | ODP | Johannesson [221] |
| | | 36-60 | ODP | Johannesson [222] |
| 1 | 3 | 1-6 | | Bussgang [64] |
| | | 17 | | Lin & Lyne [287] |
| | | 35 | | Costello [92] |
| | | 1-6 | ODP, OMD | Johannesson [223] |
| | | 7-23 | ODP | Johannesson [223] |
| 1 | 4 | 15 | | Lin & Lyne [287] |
| | | 35 | | Costello [92] |
| | | 1-23 | ODP | Johannesson [223] |
| 1 | 5-9 | 15 | | Lin & Lyne [287] |
| 2 | 3 | 1-23 | ODP | Johannesson, Paaske [224] |
| | | 9 | | Lin & Lyne [287] |
| | | 25 | ODP | Hagenauer [175] |
| 2 | 4 | 12 | ODP | Ferreira *et al.* [139] |
| 3 | 4 | 2-12 | ODP, OFD | Sun *et al.* [435] |
| | | 20 | ODP | Hagenauer [175] |
| 2 | 5 | 7 | | Lin & Lyne [287] |
| | | 12 | ODP | Ferreira *et al.* [139] |
| 3 | 5 | 12 | ODP | Ferreira *et al.* [139] |
| | | 3-8 | ODP, OFD | Sun *et al.* [435] |
| 4 | 5 | 3-13 | ODP, OFD | Sun *et al.* [435] |
| | | 20 | ODP | Hagenauer [175] |
| 3 | 6 | 6 | ODP | Ferreira *et al.* [139] |
| 5 | 6 | 3-14 | ODP, OFD | Sun *et al.* [435] |
| | | 16 | ODP | Hagenauer [175] |
| 2, 3 | 7 | 10 | ODP | Ferreira *et al.* [139] |
| 5 | 7 | 3 | ODP | Ferreira *et al.* [139] |
| 6 | 7 | 3-15 | ODP, OFD | Sun *et al.* [435] |
| | | 16 | ODP | Hagenauer [175] |
| 7 | 8 | 12 | ODP | Hagenauer [175] |

**Table A.9** References that include listings of ($n$, $k$, $m$) systematic convolutional encoders of various types.

| $k$ | $n$ | $\nu$, $m$, or $K$ | Type | Reference |
|---|---|---|---|---|
| 1 | 2 | $m = 2 - 8$ | Best $d_{free}$ | Odenwalder [345] |
| | | $m = 2 - 13$ | NC, Max $d_{free}$ | Larsen [261] |
| | | $m = 4, 11, 13$ | C, Max $d_{free}$ | Larsen [261] |
| | | $m = 2 - 6$ | QLI, Max $d_{free}$ | Helgert [192] |
| | | $m = 3 - 15$ | CMP | Chevillat & Costello [79] |
| | | $m = 2 - 23$ | CMP | Bahl & Jelinek [23] |
| | | $m = 1 - 23$ | ODP, QLI | Johannesson [221] |
| | | $m = 1 - 13$ | QLI, Max $d_{free}$ | Johannesson [221] |
| | | $m = 1 - 13$ | ODP, OMD, OFD | Johannesson [221] |
| | | $m = 24 - 50$ | ODP, QLI | Johannesson [222] |
| | | $m = 1 - 23$ | ODP | Johannesson, Paaske [224] |
| | | $m =$ 11,12, 14,16 | OFD | Johannesson, Paaske [224] |
| | | $m = 3 - 23$ | ELIT | Johannesson, Paaske [224] |
| | | $m =$ 4,8, $15 - 17$,21,23 | ELI | Johannesson, Paaske [224] |
| 1 | 3 | $m = 2 - 7$ | Best $d_{free}$ | Odenwalder [345] |
| | | $m = 2 - 13$ | NC, Max $d_{free}$ | Larsen [261] |
| | | $m = 1 - 7$ | ODP, OFD | Johannesson [223] |
| 1 | 4 | $m = 2 - 13$ | NC, Max $d_{free}$ | Larsen [261] |
| 1 | $5 - 8$ | $m = 2 - 7$ | Best $d_{free}$ | Daut *et al.* [100] |
| 2 | 3 | $K = 3 - 8$ | Best $d_{free}$ | Daut *et al.* [100] |
| | | $\nu = 2 - 10$ | | Paaske [354] |
| | | $m = 3 - 23$ | ODP | Johannesson, Paaske [224] |
| 3 | 4 | $K = 4 - 8$ | Best $d_{free}$ | Daut *et al.* [100] |
| | | $\nu = 2 - 10$ | | Paaske [354] |
| 2 | 5, 7 | $K = 3 - 8$ | Best $d_{free}$ | Daut *et al.* [100] |
| 3 | 5,7,8 | $K = 4 - 8$ | Best $d_{free}$ | Daut *et al.* [100] |
| 4 | 5,7 | $K = 5 - 8$ | Best $d_{free}$ | Daut *et al.* [100] |
| 5 | $6 - 8$ | $K = 6 - 8$ | Best $d_{free}$ | Daut *et al.* [100] |
| 6 | 7 | $K = 7, 8$ | Best $d_{free}$ | Daut *et al.* [100] |
| 7 | 8 | $K = 8$ | Best $d_{free}$ | Daut *et al.* [100] |

**Table A.10** References that include listings of $(n, k, m)$ nonsystematic convolutional encoders of various types.

| Class | Rates | Reference |
|---|---|---|
| Convolutional self-orthogonal codes (CSOC) | $(n-1)/n$, $n$ =2-23, 25 26, 28, 29, 31, 32, 35 37, 38, 41-43, 46, 50 | Wu [476, 477, 478] |
| Min SNR for given BER with Viterbi decoding | | P.J. Lee [271, 272] |
| Unit Memory Codes | | L.N. Lee [269]<br>Lauer [263]<br>Justesen *et al.* [230]<br>Said & Palazzo [391] |
| Punctured codes | | Cain *et al.* [68]<br>Yasuda *et al.* [487]<br>P.J. Lee [273]<br>Hole [205]<br>Haccoun & Begin [171, 26]<br>Begin *et al.* [27] |
| RCPC codes | | Hagenauer *et al.* [178, 178] |
| RCC codes | | Kallel & Haccoun [241] |

**Table A.11** References that include listings of some specific classes of convolutional encoders (RCPC = Rate-compatible punctured convolutional, RCC = Rate-compatible convolutional).

| Property | Rates | Reference |
|---|---|---|
| Distance Spectra | 1/2 | Cedervall & Johannesson [70] |
| | 1/2, 1/3, 1/4 | Conan [89] |
| Decision depths | 1/3, 1/2, 2/3, 3/4 | Anderson & Balachandran [15, 13] |

**Table A.12** References that include listings of some properties of convolutional encoders.

# PART III

# ERROR CORRECTION AND DECODING

# 5

# VITERBI DECODING

Decoding of convolutional codes is the focus of Part III of this book. Chapters 5 through 8 describe four different decoding schemes, namely, sequential, majority-logic, Viterbi, and table look-up decoding. Viterbi decoding is optimal in the maximum likelihood sense and has become a natural reference for comparing other decoding schemes. Consequently, we begin our discussion with Viterbi decoding.

## 5.1 MAXIMUM LIKELIHOOD DECODING

In the block diagram of a digital communication system shown in Figure 5.1, the source information sequence $\mathbf{u}$ is fed to an encoder which produces the encoded sequence $\mathbf{v}$ as its output. This sequence is transmitted over a digital channel. The received sequence is denoted by $\mathbf{r}$. The channel noise may introduce errors in the received sequence, and hence $\mathbf{r}$ may not be the same as $\mathbf{v}$. The error sequence is denoted by $\mathbf{e}$ and we have $\mathbf{r} = \mathbf{v} + \mathbf{e}$. At the receiver, a decoder attempts to recover the original source information from the received sequence. Let the estimated information sequence be $\hat{\mathbf{u}}$. The decoder may directly recover $\hat{\mathbf{u}}$ from $\mathbf{r}$ or it may first obtain the estimated encoded sequence $\hat{\mathbf{v}}$ and use it to get $\hat{\mathbf{u}}$.

The conversion of $\mathbf{r}$ into $\hat{\mathbf{v}}$ is called *error correction* and that of $\hat{\mathbf{v}}$ into $\hat{\mathbf{u}}$ is called *inversion*. The error correction is sometimes referred to as codeword estimation. Once a codeword estimate is available, the task of obtaining the estimated information sequence is simple; all that needs to be done is an inverse mapping from an encoded sequence to an information sequence. Often, it is

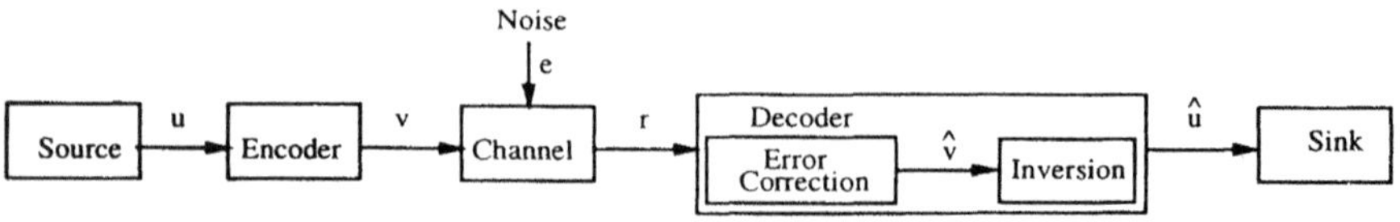


**Figure 5.1** A digital communication system.

possible to obtain $\hat{\mathbf{u}}$ directly from $\mathbf{r}$. That is, the error correction and the inversion stages are not separate. It is a common practice to refer to either type of conversion carried out at the receiver as *decoding*.

A reasonable decoding rule is to decode each received word into the codeword "closest" to it; this is called *nearest-neighbor decoding*. If all codewords are equally likely and if the decoding scheme estimates $\hat{\mathbf{v}}$, given the sequence $\mathbf{r}$, by minimizing the probability that $\hat{\mathbf{v}}$ and $\mathbf{v}$ differ, i.e., $\hat{\mathbf{v}}$ is the most likely codeword given that $\mathbf{r}$ was received, it is called *maximum likelihood decoding*. Given channel transition probabilities below 0.5, the maximum likelihood decoding for a BSC is equivalent to selection of the codeword which minimizes the Hamming distance between $\mathbf{r}$ and $\mathbf{v}$.

## 5.2 THE VITERBI ALGORITHM

The Viterbi algorithm was proposed by Viterbi [451] in 1967 as an asymptotically optimum decoding technique for convolutional codes. The algorithm was shown to be equivalent to dynamic programming, and hence optimal, by Omura [350]. Forney recognized that the Viterbi algorithm is in fact always optimum in the maximum likelihood sense [150, 151]. Viterbi further clarified his algorithm in the context of convolutional coding [452]. It has become the most commonly used convolutional decoding algorithm and has found applications in many other areas [150].

### 5.2.1 A General Version

Forney [150] showed that the Viterbi algorithm solves a more general problem which can be stated as follows: Given a sequence $\hat{\mathbf{v}}$ of observations of a discrete-state Markov process in memoryless noise, find the state sequence $\mathbf{x}$ for which the *a posteriori* probability $P(\mathbf{x}|\hat{\mathbf{v}})$ is maximum.

In the underlying Markov process, time is discrete. The state $x_t$ at time $t$ is one of a finite number $B$ of states $b$ from the state space $X = \{0, 1, \ldots, B-1\}$. We shall assume that the process runs from time 0 to time $T$, and that the initial and final states $x_0$ and $x_T$ are known. The finite vector $\mathbf{x} = (x_0, \ldots, x_T)$ then represents the state sequence.

The process is Markov in the sense that the probability of being in state $x_{t+1}$ at time $t+1$, given all states up to time $t$, depends only on the state $x_t$ at time $t$, i.e., $P(x_{t+1}|x_0, x_1, \ldots, x_t) = P(x_{t+1}|x_t)$. The transition from a state $x_t$ to another state $x_{t+1}$ is $\xi_t = (x_t, x_{t+1})$. There is a one-to-one correspondence between state sequence $\mathbf{x}$ and the transition sequence $\xi = (\xi_0, \ldots, \xi_{T-1})$. The noise associated with the process is assumed to be memoryless. Then there is a sequence $\hat{\mathbf{v}}$ of observations $\hat{v}_t$ in which $\hat{v}_t$ depends probabilistically only on the transition $\xi_t$ at time $t$:

$$P(\hat{\mathbf{v}}|\mathbf{x}) = P(\hat{\mathbf{v}}|\xi) = \prod_{t=0}^{T-1} P(\hat{v}_t|\xi_t).$$

Consider, for example, the following model. Let $\mathbf{u} = (u_0, u_1, \ldots)$ be an input sequence. Each $u_t$ is generated independently according to some probability distribution. There is a noise-free signal sequence $\mathbf{v}$, not observable, in which each $v_t = f(u_t, \ldots, u_{t-\nu})$ is some deterministic function of the present and $\nu$ previous inputs. This sequence is fed into a memoryless channel whose output is the observed sequence $\hat{\mathbf{v}}$. Such a process is called a *shift-register process*. It can be modeled by a shift register of length $\nu$ with inputs $u_t$. The state of the shift-register process is $x_t = (u_{t-1}, \ldots, u_{t-\nu})$. If $\mathbf{u}$ is a binary sequence, we have $B = 2^\nu$. The transition from a state $x_t$ to another state $x_{t+1}$ is $\xi_t = (u_t, \ldots, u_{t-\nu})$. Given a sequence of observations $\hat{\mathbf{v}}$, every path through the encoder trellis can be assigned a length proportional to $-\ln P(\mathbf{x}, \hat{\mathbf{v}})$, where $\mathbf{x}$ is the state sequence associated with that path. Thus the problem of finding the state sequence for which $P(\mathbf{x}|\hat{\mathbf{v}})$ is maximum, or equivalently for which $P(\mathbf{x}, \hat{\mathbf{v}}) = P(\mathbf{x}|\hat{\mathbf{v}})P(\hat{\mathbf{v}})$ is maximum, can be solved by finding the path whose length $-\ln P(\mathbf{x}, \hat{\mathbf{v}})$ is minimum, since $\ln P(\mathbf{x}, \hat{\mathbf{v}})$ is a monotonic function of $P(\mathbf{x}, \hat{\mathbf{v}})$ and there is a one-to-one correspondence between paths and sequences. Therefore

$$\begin{aligned} P(\mathbf{x}, \hat{\mathbf{v}}) &= P(\mathbf{x})P(\hat{\mathbf{v}}|\mathbf{x}) \\ &= \prod_{t=0}^{T-1} P(x_{t+1}|x_t) \prod_{t=0}^{T-1} P(\hat{v}_t|x_{t+1}, x_t) \end{aligned}$$

and the length assigned to each branch is the following:

$$\lambda(\xi_t) = -\ln P(x_{t+1}|x_t) - \ln P(\hat{v}_t|\xi_t).$$

The total length of the path corresponding to some $\mathbf{x}$ is

$$-\ln P(\mathbf{x}, \hat{\mathbf{v}}) = \sum_{t=0}^{T-1} \lambda(\xi_t).$$

Let us denote a segment $(x_0, x_1, ..., x_t)$ of the state sequence $\mathbf{x}$ by $\mathbf{x}_0^t$. It consists of states up to time $t$. In the encoder trellis $\mathbf{x}_0^t$ corresponds to a path segment starting at the node $x_0$ and terminating at $x_t$, with length

$$\lambda(\mathbf{x}_0^t) = \sum_{i=0}^{t-1} \lambda(\xi_i).$$

For a node $x_t$ in the trellis, there may be several path segments leading up to it. The shortest such path segment is called the *survivor* corresponding to node $x_t$, and is denoted $\hat{\mathbf{x}}(x_t)$. At any time $t$ we need to remember the $B$ survivors $\hat{\mathbf{x}}(x_t)$ and their lengths $\Gamma(x_t) = \lambda[\hat{\mathbf{x}}(x_t)]$. The decoding algorithm is given below.

*Storage:* Memory is needed for the following:

1. $t$ (time index)
2. $\hat{\mathbf{x}}(x_t)$, $1 \leq x_t \leq B$ (survivor terminating in $x_t$)
3. $\Gamma(x_t)$, $1 \leq x_t \leq B$ (survivor length).

*Initialization:* Assuming the starting state to be $x_0$, set

1. $t = 0$;
2. $\hat{\mathbf{x}}(x_0) = x_0$; $\hat{\mathbf{x}}(b)$ arbitrary, $b \neq x_0$;
3. $\Gamma(x_0) = 0$; $\Gamma(b) = \infty, b \neq x_0$.

*Recursion:* Repeat the following steps until $t = T$:

**Step 1.** Compute

$$\Gamma(x_{t+1}, x_t) = \Gamma(x_t) + \lambda[\xi_t = (x_{t+1}, x_t)]$$

for all $\xi_t = (x_{t+1}, x_t)$.

**Step 2.** Find

$$\Gamma(x_{t+1}) = \min_{x_t} \Gamma(x_{t+1}, x_t)$$

for each $x_{t+1}$; store $\Gamma(x_{t+1})$ and corresponding survivor $\hat{\mathbf{x}}(x_{t+1})$.

**Step 3.** Set $t$ to $t+1$ and go back to step 1.

*Result:* The shortest path is the survivor $\hat{\mathbf{x}}(x_T)$.

### 5.2.2 An Example Run

The operation of the Viterbi algorithm can be explained using the encoder trellis diagram described in Chapter 2. Figure 5.2 shows the trellis diagram of the (2, 1, 2) encoder with generator sequences $\mathbf{g}^{(1)} = 101$ and $\mathbf{g}^{(2)} = 111$. The four states are marked 00, 01, 10, and 11, and the time is marked horizontally. The initial state is assumed to be 00. In the trellis, the upper outgoing branch for each state corresponds to an input of 0, whereas the lower outgoing branch corresponds to an input of 1. Each state also has two incoming branches. The encoder outputs are indicated for each branch in the trellis.

Each information sequence, uniquely encoded into an encoded sequence, corresponds to a unique path in the trellis. Equivalently, given a path through the trellis, the corresponding encoded sequence (information sequence ) can be obtained by reading off the output (input) labels on all the branches that make up the path. Consider the information sequence $\mathbf{u} = 1011101$ of length $N = 7$ bits. The corresponding encoded sequence, obtained after appending $m = 2$ terminating zeros to $\mathbf{u}$, is $\mathbf{v} = 11\ 01\ 00\ 10\ 01\ 10\ 00\ 01\ 11$. The path through the trellis corresponding to this sequence is shown highlighted in Figure 5.2.

The received sequence can potentially be erroneous, and therefore may not exactly correspond to a path in the trellis. Viterbi algorithm operates by computing a log-likelihood metric for each path at each level of the trellis. This metric is calculated and maintained as a partial metric for each branch as the trellis is traversed. For a BSC, the log-likelihood metric is the Hamming weight of the output corresponding to a branch. Since there are two paths merging at each node, the path with a smaller metric is selected while the other path is discarded, thereby keeping only one surviving path per state per time unit. The surviving paths and the corresponding metrics are stored in memory. The received sequence ends with a series of all zero input bits. This brings the number of surviving paths down to one, which is the path with the smallest distance from the received sequence.

Consider the received sequence $\mathbf{r} = 11\ 01\ 00\ 11\ 11\ 10\ 00\ 01\ 11$. The Viterbi algorithm begins by comparing the first received branch 11 in the first time unit to the first two possible code branches, namely, 00 and 11. Note that we have assumed the all-zero state as the starting state, so that all the states are reached only after $m$ time units. The distance of the code branch 00 from received branch 11 is 2, while that of 11 is 0. This distance, or metric, is stored in the memory for the two states 00 and 01. The metric for each state at every time unit is shown in Figure 5.2. At time unit 1, there are only two paths.

These are expanded to four paths at time unit 2. The four paths now reach all the four states. The distance of the second received branch 01 from the second code branches on all the four paths is computed. It is called the branch metric. The branch metrics are added to the path metrics of each state at time unit 1 to obtain the path metrics of each state for time unit 2. These are labeled as 3, 3, 0, and 2 for states 00, 01, 10, and 11, respectively, in Figure 5.2.

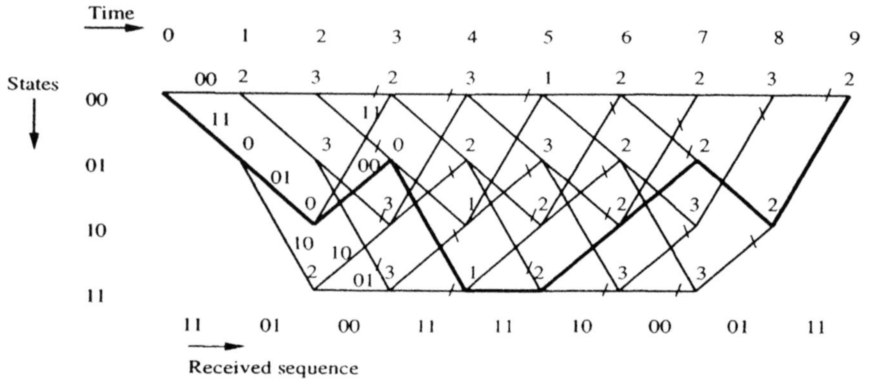


**Figure 5.2** Viterbi algorithm applied to a rate one-half encoder trellis.

At time unit 3, each state has two incoming paths. The branch metrics of each of the two incoming branches are computed and added to the path metrics of the previous states of these branches. The path with the lower metric is retained while the one with higher metric is discarded. The retained path for each state is called the *survivor*. For instance, consider state 00. The two incoming branches are the one with output label 00 from state 00 and the one with output label 11 from state 10. The branch metrics for these two branches are 0 and 2, respectively, since the received branch at this time unit is 00. Adding these to the path metrics 3 (at state 00) and 0 (at state 10), we get path metrics 3 and 2, respectively. The path with metric 3 is discarded and the survivor is the path with metric 2. This process is repeated for all the states. In case of a tie, one of the two paths is arbitrarily discarded. In our example, we discard the lower incoming path whenever there is a tie. In Figure 5.2, the discarded paths are marked with a '/' or a '\' at the discarded branch.

This process is continued for time units 4 though 7. Time units 8 and 9 correspond to the terminating 0's, and so only the branches with 0 input are considered. This reduces the number of surviving paths to two and eventually to one. The survivor in this case has a metric equal to 2, indicating that the

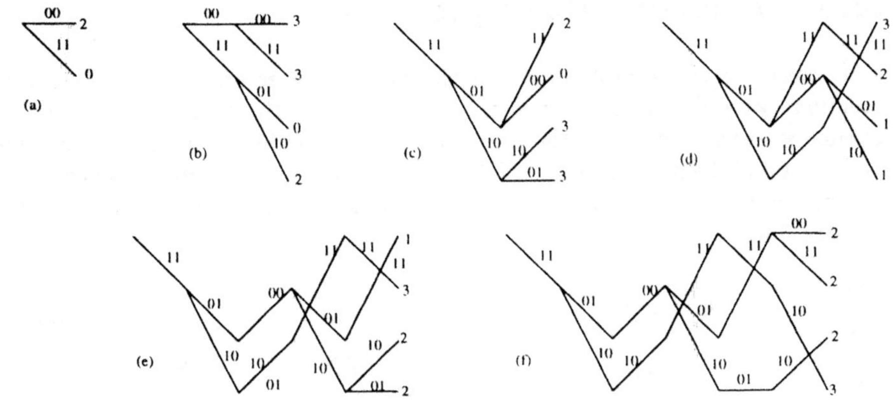


**Figure 5.3** Stage by stage progress of the Viterbi algorithm (stages (a) - (f)).

received sequence **r** has two errors. The surviving path is shown highlighted in Figure 5.2. The estimated encoded sequence corresponding to the survivor is obtained by reading off the output labels on all the branches of the path and is found to be $\hat{\mathbf{v}}$ = 11 01 00 10 01 10 00 01 11, which is the same as **v**. The estimated information sequence $\hat{\mathbf{u}}$ is obtained by noting whether each branch in the surviving path corresponds to an input of 0 or 1. The progress of the Viterbi algorithm at each time unit is shown in Figure 5.3. Only the survivor paths are shown.

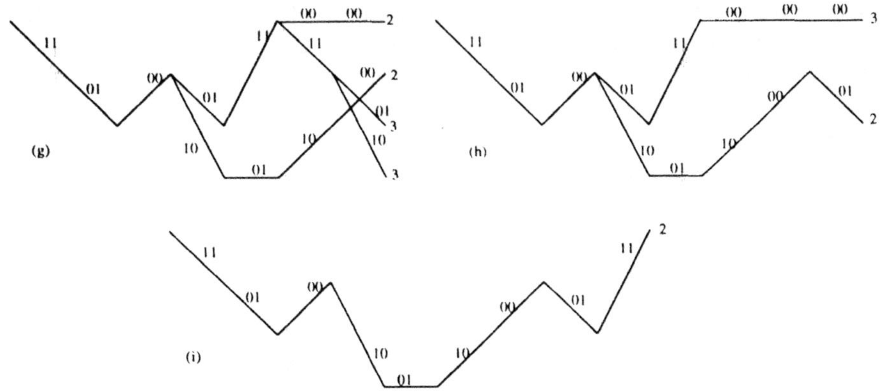


**Figure 5.3** *(continued)* Stages (g) - (i).

### 5.2.3 Another Example Run

Let us consider an example in which the number of errors exceeds the error-correcting capability $t_{free} = 2$ of the encoder. Let the transmitted encoded sequence $\mathbf{v}$ be the same as in the previous section, and let the received sequence be $\mathbf{r}$ = 11 00 01 11 01 10 00 01 11. The Viterbi algorithm proceeds as described in the previous section. The progress of the algorithm at each time unit is shown in Figure 5.4.

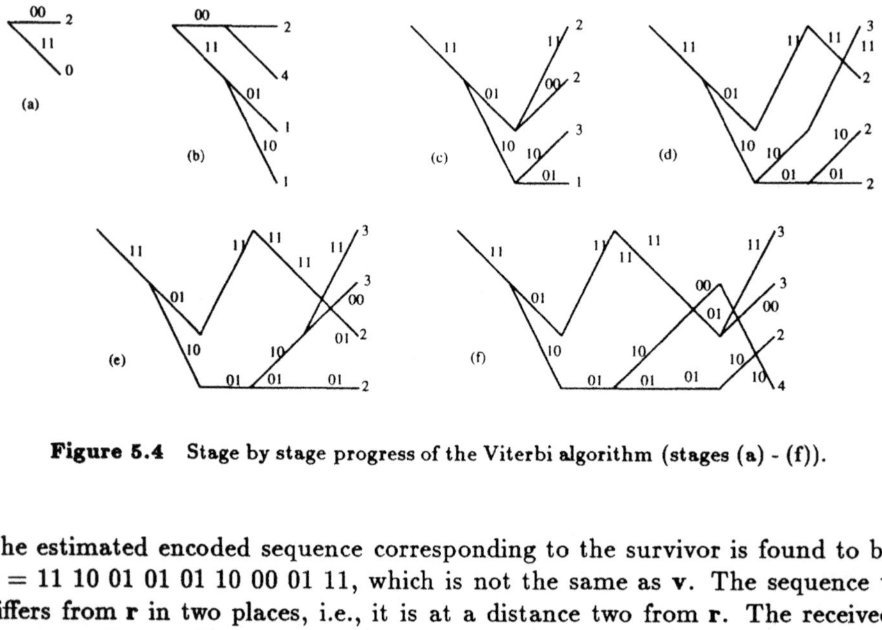

**Figure 5.4** Stage by stage progress of the Viterbi algorithm (stages (a) - (f)).

The estimated encoded sequence corresponding to the survivor is found to be $\hat{\mathbf{v}}$ = 11 10 01 01 01 10 00 01 11, which is not the same as $\mathbf{v}$. The sequence $\hat{\mathbf{v}}$ differs from $\mathbf{r}$ in two places, i.e., it is at a distance two from $\mathbf{r}$. The received sequence $\mathbf{r}$ is at a distance three from $\mathbf{v}$, and the corresponding error sequence is $\mathbf{e}$ = 00 01 01 01 00 00 00 00 00. Thus, in this example, a three bit error made the received sequence closer in distance to the encoded sequence $\hat{\mathbf{v}}$ than the encoded sequence $\mathbf{v}$. The Viterbi algorithm correctly estimated $\hat{\mathbf{v}}$ as the "nearest-neighbor" encoded sequence based on $\mathbf{r}$.

## 5.3 PERFORMANCE ANALYSIS

The performance of Viterbi decoding can be evaluated using the transfer function $T(X)$ of the encoder. Assume for the moment that a BSC is being used. Without loss of generality, it is assumed that the all-zero code sequence $\mathbf{0}$ is

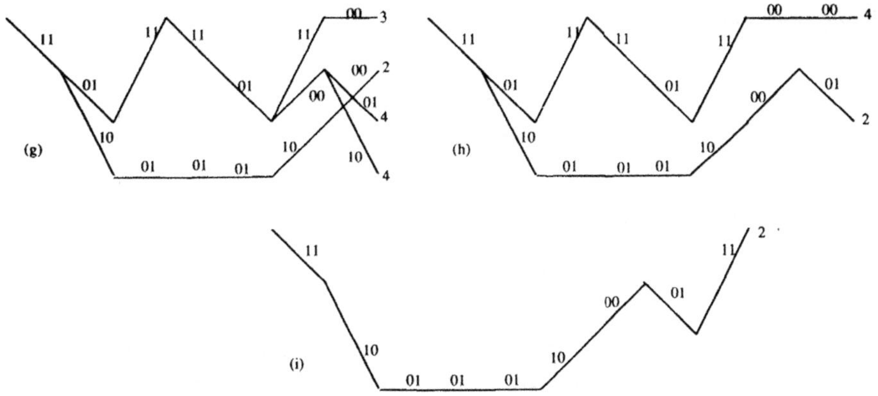


**Figure 5.4** *(continued)* Stages (g) - (i).

transmitted. Thus, errors in the received sequence show up as 1's. The decoder makes an error if the *correct* (all-zero) path in the trellis is discarded at some time unit $j$ in favor of a competing (*incorrect*) path. Consider the first such event in decoding a received sequence. It is known as the *first event error*. If the incorrect path has weight $d$, a first event error on a BSC with transition probability $p$ is made with probability

$$P_d = \begin{cases} \displaystyle\sum_{e=(d+2)/2}^{d} \binom{d}{e} p^e(1-p)^{d-e}, & d \text{ odd} \\ \displaystyle\frac{1}{2}\binom{d}{d/2} p^{d/2}(1-p)^{d/2} + \sum_{e=(d/2)+1}^{d} \binom{d}{e} p^e(1-p)^{d-e}, & d \text{ even.} \end{cases} \quad (5.1)$$

The factor of $\frac{1}{2}$ in the first term when $d$ is even is the probability of an error if the metrics of the correct and incorrect paths are tied. Considering all incorrect paths, the first event error probability at time unit $j$, $P_f(E,j)$, is overbounded by

$$P_f(E,j) < \sum_{d=d_{free}}^{\infty} A_d P_d, \quad (5.2)$$

where $A_d = n(d)$ is the number of encoded sequences of weight $d$. Since this bound is independent of $j$, the *first event error probability* at any time unit, $P_f(E)$, is bounded by

$$P_f(E) < \sum_{d=d_{free}}^{\infty} A_d P_d. \quad (5.3)$$

Using the Chernoff bound $P_d < [2\sqrt{p(1-p)}]^d$ [452, 446], we get

$$P_f(E) < \sum_{d=d_{free}}^{\infty} A_d[2\sqrt{p(1-p)}]^d. \tag{5.4}$$

For an arbitrary convolutional encoder with a generating function $T(X) = \sum_{d=d_{free}}^{\infty} A_d X^d$, (5.4) can be rewritten as

$$P_f(E) < T(X)|_{X=2\sqrt{p(1-p)}}. \tag{5.5}$$

More event errors may occur after the first event error as the decoding progresses. It can be shown that the *event error probability* (not just the first) at any time unit, $P(E)$, is bounded by

$$P(E) < \sum_{d=d_{free}}^{\infty} A_d P_d < T(X)|_{X=2\sqrt{p(1-p)}}. \tag{5.6}$$

For small values of $p$, the free distance term dominates the bound, and the event error probability is approximately

$$P(E) \approx A_{d_{free}}[2\sqrt{p(1-p)}]^{d_{free}} \tag{5.7}$$

$$\approx A_{d_{free}} 2^{d_{free}} p^{d_{free}/2} \approx A_{d_{free}} 2^{d_{free}} p^{t_{free}/2}. \tag{5.8}$$

The *bit error probability* $P_b(E)$ is the expected number of information bit decoding errors per decoded information bit. In terms of the transfer function, $P_b(E)$ is upper bounded by

$$P_b(E) \leq \frac{1}{k} \frac{\partial\, T(X,Y)}{\partial\, Y}|_{X=2\sqrt{p(1-p)}, Y=1}. \tag{5.9}$$

In general, on a binary input, $Q$-ary output DMC, we have

$$P_b(E) \leq \frac{1}{k} \frac{\partial\, T(X,Y)}{\partial\, Y}|_{X=\sum_{j=0}^{Q-1}\sqrt{p(j|0)p(j|1)}, Y=1}, \tag{5.10}$$

where $p(j|0)$ and $p(j|1)$ are the probabilities that the decoder output will be the quantized level $j$ when the transmitted symbol is 0 and 1, respectively.

The transfer function

$$T(X,Y) = \sum_{d=d_{free}}^{\infty} \sum_{j=0}^{\infty} A_{d,j} X^d Y^j \tag{5.11}$$

when differentiated with respect to $Y$ and evaluated at $Y = 1$ gives

$$\frac{\partial\, T(X,Y)}{\partial\, Y}\Big|_{Y=1} = \sum_{d=d_{free}}^{\infty} \sum_{j=0}^{\infty} j A_{d,j} X^d \tag{5.12}$$

$$= \sum_{d=d_{free}}^{\infty} B_d X^d, \tag{5.13}$$

where $B_d = \sum_{j=1}^{\infty} j A_{d,j} = n_b(d)$ is the total number of nonzero information bits on all weight $d$ paths. Hence, the bit error probability is given by

$$P_b(E) < \frac{1}{k} \sum_{d=d_{free}}^{\infty} B_d P_d. \tag{5.14}$$

Again, for small values of $p$, the free distance term dominates the bound, and the bit error probability is approximately

$$P_b(E) \approx \frac{1}{k} B_{d_{free}} [2\sqrt{p(1-p)}]^{d_{free}} \tag{5.15}$$

$$\approx \frac{1}{k} B_{d_{free}} 2^{d_{free}} p^{d_{free}/2} \approx \frac{1}{k} B_{d_{free}} 2^{d_{free}} p^{t_{free}/2}. \tag{5.16}$$

The bit error probability given above is an upper bound obtained by Viterbi [452]. Tighter upper bounds for the BSC case are given by van de Meeberg [446] and Post [364].

If the BSC is implemented by an AWGN channel with BPSK modulation, coherent detection, and hard-decision decoding, the transition probability $p$ is given by

$$p = \frac{1}{2} \mathrm{erfc}\left(\sqrt{\frac{\mathrm{RE_b}}{\mathrm{N_0}}}\right) \approx \frac{1}{2} \mathrm{e}^{-\mathrm{RE_b/N_0}}, \tag{5.17}$$

where $R$ is the coding rate, $E_b$ is the energy per information bit, and $N_0$ is the one-sided noise power spectral density. Substituting (5.17) in (5.16), we get

$$P_b(E) \approx \frac{1}{k} B_{d_{free}} 2^{d_{free}/2} e^{-R d_{free} E_b / 2N_0}. \tag{5.18}$$

The bit error probability without coding ($R = 1$) is

$$P_b(E) \approx \frac{1}{2} e^{-E_b/N_0}. \tag{5.19}$$

Notice that the exponent in (5.18) is $Rd_{free}/2$ times that in (5.19). Thus, in order to achieve the same asymptotic error probability as in the coded case, the ratio $E_b/N_0$ in the uncoded case must be increased by a factor of $Rd_{free}/2$. This factor is called the *asymptotic coding gain* $\gamma$, and is expressed in decibels as

$$\gamma = 10\log_{10}\left(\frac{Rd_{free}}{2}\right). \tag{5.20}$$

This is the coding gain achievable when hard decision decoding is used. The asymptotic coding gain for an unquantized channel outputs is 3 dB more than $\gamma$, and soft decision decoding with $Q = 8$ levels achieves most of this 3 dB gain.

The Viterbi algorithm performs a fixed number of calculation per time unit even if there are no errors in a given received sequence. Furthermore, the size of the trellis (number of states) doubles for each additional memory element in the encoder when binary data is used. Encoders of constraint length larger than 8 require very complex decoder implementation.

## 5.3.1 Truncated Viterbi Decoding

A truly maximum likelihood decoder must examine the entire received sequence before making any decoding decisions. This means that the Viterbi algorithm, in principle, has to be able to store the survivor paths and their metrics until the terminating $m$ stages are processed. For an information sequence of length $M$ blocks (or an encoded sequence of length $N = M + m$ blocks), this may result in excessive storage requirements. In practice, of course, only a finite amount of memory is used. Only the most recent $\tau$ blocks are stored for each survivor, where $\tau \ll M$. This approach is known as path memory truncation, and the resulting decoding scheme is often called *truncated* Viterbi decoding [199, 351].

Once the decoding has progressed up to the $\tau$th level in the trellis, the path memory is full. A final decoding decision has to be made on the first information block before a new block can be added to each survivor path. In general, a decoding decision has to be made on $(t - \tau)$th information block at time $t$. The decoding decision may involve selecting the $(t - \tau)$th information block corresponding to the best survivor at time $t$, or selecting the information block that appears most often in all the survivors, or simply selecting the $(t - \tau)$th information block of an arbitrary survivor. Note that the decoding decisions made this way are not maximum likelihood. The decoded information block may come from a survivor path that may get rejected later in favor of another path at some later stage. However, the truncated Viterbi decoder can be made very close to a maximum likelihood decoder by selecting $\tau$ sufficiently large.

It has been observed that for rate one-half encoders, all the survivors at time $t$ stem from the same information block at time $(t - \tau)$ with probability very close to 1 if $\tau$ is chosen to be five to six times the memory order $m$ or larger [151, 82].

### 5.3.2 Effect of Soft Decisions

The performance of Viterbi decoding can be improved by using a soft quantized demodulator output. The channel in this case is modeled as a binary input, $Q$-ary output DMC with $Q > 2$. For $Q = 2$, the DMC reduces to a BSC. An ideal DMC uses infinite quantization, i.e., has an unquantized analog output. In practice, however, typical values of $Q$ that are used are 8, 16, etc. It has been shown that with $Q = 8$, the performance improvement is very close to that achievable with unquantized demodulator output [197].

For each input bit, the output of a soft decision DMC with $Q > 2$ is a $Q$-ary symbol. For instance, the outputs for $Q = 4$ can be labeled as $0_1$, $0_2$, $1_2$, and $1_1$. The symbols $0_1$ and $1_1$ are in a sense "stronger" than the symbols $0_2$ and $1_2$, respectively. On a BSC, both $0_1$ and $0_2$ would be seen as just 0, and both $1_1$ and $1_2$ as 1. So, the 4-ary DMC provides more information about the channel noise level. A symmetric DMC with $Q = 4$ is shown in Figure 5.5.

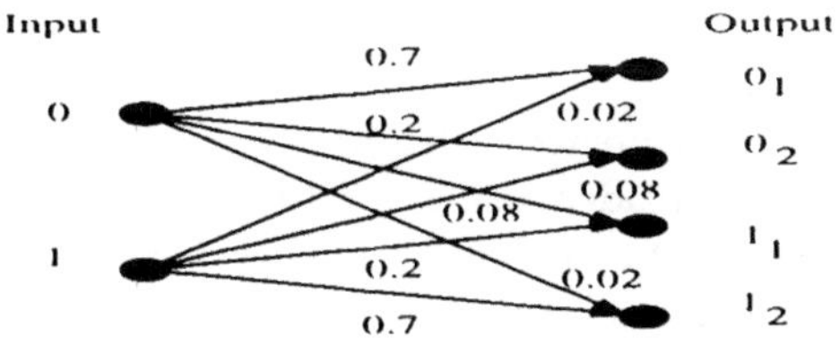


**Figure 5.5** A binary input, 4-ary output symmetric DMC.

A maximum-likelihood decoder maximizes the log-likelihood function log $P(\mathbf{r}|\mathbf{v})$, i.e., it maximizes the probability of getting a received sequence $\mathbf{r}$ given the transmitted encoded sequence $\mathbf{v}$. Assume that an information sequence $\mathbf{u}$ of length $N$ blocks is encoded using an $(n, k, m)$ encoder to get an encoded sequence $\mathbf{v}$ of length $M = n(N + m)$ blocks, which is sent over a binary input, $Q$-ary output DMC. The corresponding $Q$-ary received sequence is $\mathbf{r}$. For convenience, let the encoded and received sequences be $\mathbf{v} = (v_0, v_1, \ldots, v_{M-1})$

and $\mathbf{r} = (r_0, r_1, \ldots, r_{M-1})$, respectively, in which the subscripts represent the ordering of the symbols.

The log-likelihood function is given by

$$\log P(\mathbf{r} \mid \mathbf{v}) = \log \prod_{i=1}^{M} P(r_i \mid v_i) = \sum_{i=1}^{M} \log P(r_i \mid v_i). \tag{5.21}$$

The log-likelihood metric table for the 4-ary DMC of Figure 5.5 is shown in Table 5.1. These real numbers can be converted to integer metrics by using the

| | $r_i$ (real numbers) | | | | $r_i$ (integers) | | | |
|---|---|---|---|---|---|---|---|---|
| $v_i$ | $0_1$ | $0_2$ | $1_2$ | $1_1$ | $0_1$ | $0_2$ | $1_2$ | $1_1$ |
| 0 | −0.16 | −0.7 | −1.1 | −1.7 | 15 | 10 | 6 | 0 |
| 1 | −1.7 | −1.1 | −0.7 | −0.16 | 0 | 6 | 10 | 15 |

**Table 5.1** Log-likelihood metrics for a binary input, 4-ary output DMC.

following expression: $c_2[\log P(r_i \mid v_i) + c_1]$, where $c_1$ is a real number chosen to make the smallest metric zero and $c_2$ is a positive real number chosen to make the metrics approximately equal to integers. The integer metrics for the DMC of Figure 5.5, obtained by choosing $c_1 = 1.7$ and $c_2 = 10$, are also given in Table 5.1.

Consider an information sequence of length $N = 5$ given by $\mathbf{u} = 10111$. This information sequence is encoded using the (2, 1, 2) encoder with $\mathbf{g}^{(1)} = 101$ and $\mathbf{g}^{(2)} = 111$ to obtain the encoded sequence $\mathbf{v} = 11\ 01\ 00\ 10\ 01\ 10\ 11$ of length $M = 2(N+m) = 14$. The sequence $\mathbf{v}$ is transmitted over the DMC and is received as the 4-ary sequence

$$\mathbf{r}_{DMC} = 1_1 1_1\ 0_1 1_1\ 0_2 1_2\ 1_2 1_2\ 0_1 0_2\ 1_1 0_1\ 1_1 1_1.$$

Note that on a BSC, the received sequence would be $\mathbf{r}_{BSC} = 11\ 01\ 01\ 11\ 00\ 10\ 11$, and the corresponding estimated encoded sequence obtained using hard decision Viterbi decoding would be $\hat{\mathbf{v}}_{BSC} = 11\ 01\ 11\ 11\ 10\ 10\ 11$. The sequence $\hat{\mathbf{v}}_{BSC}$ corresponds to the estimated information sequence $\hat{\mathbf{u}}_{BSC} = 10011$, which is not equal to the original information sequence $\mathbf{u}$. Thus, decoding errors are said to have occured.

The outcome of using the Viterbi algorithm on the soft quantized received sequence $\mathbf{r}_{DMC}$ with the integer metrics given in Table 5.1 is shown in Figure

5.6. The estimated encoded sequence is

$$\hat{\mathbf{v}}_{DMC} = 11\ 01\ 00\ 10\ 01\ 10\ 11,$$

and the corresponding estimated information sequence is $\hat{\mathbf{u}}_{BSC} = 10111$, which is equal to the original information sequence $\mathbf{u}$. Thus, soft decision Viterbi decoding results in no decoding errors.

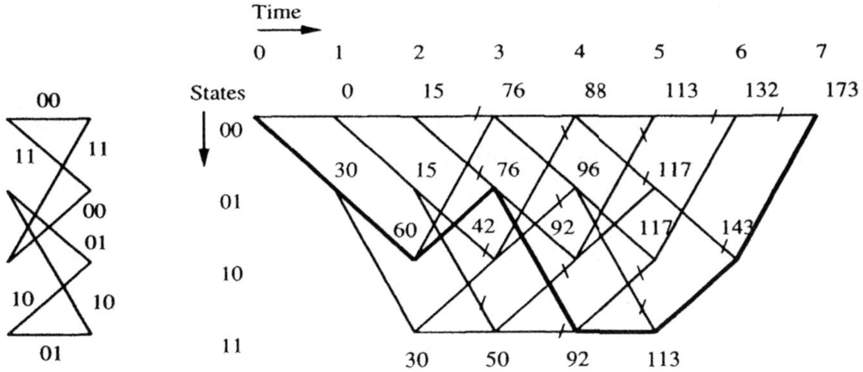


**Figure 5.6** Viterbi decoding with soft decisions on a (2, 1, 2) encoder trellis.

## 5.4 ENCODER SELECTION

The performance of Viterbi decoding is determined by the free distance and the distance spectrum of the code or the encoder, as seen in the previous section. Thus, given a coding rate, an encoder with as large a value of $d_{free}$ as possible should be selected, with $A_{d_{free}}$ and $B_{d_{free}}$ determining the choice among the codes with the same $d_{free}$. From Section 4.12, we see that larger free distance can be obtained by selecting encoders with larger constraint lengths. However, the constraint length of an encoder of a given rate cannot be increased indefinitely since the resulting Viterbi decoder has very large computation and storage requirements. For listings of encoders with best $d_{free}$, see Section 4.12.

## 5.5 OTHER ALGORITHMS RELATED TO VITERBI DECODING

### 5.5.1 Viterbi Decoding of Punctured Codes

Punctured codes and encoders were discussed in Section 4.8. It was shown that many high-rate punctured encoders can be obtained from the same low-rate original encoder. This fact can be exploited in devising a variable-rate Viterbi decoder. The advantage of using punctured codes is that a high-rate punctured code can be decoded using the decoder for the low-rate original code, thereby requiring a smaller number of computations. In Example 4.11, the equivalent ordinary rate two-thirds encoder trellis with $k = 2$ has $2^k = 4$ branches entering and leaving each state. Viterbi decoding using this trellis requires a 4-ary comparison per state for every two information bits. A 4-ary comparison is implemented as three binary comparisons in a serial implementation. Hence, the ordinary encoder requires one and half binary comparisons per state for each information bit. By contrast, the punctured encoder requires one binary comparison per state for each information bit. Therefore, using the punctured encoder has some speed advantages. Cain *et al.* [68] described the Viterbi decoding of punctured codes, and Yasuda *et al.* [486, 487] described an optimum soft decision Viterbi decoding scheme and a variable-rate Viterbi decoder. For more details, see [171, 26, 27].

### 5.5.2 Syndrome Decoding

A technique called *syndrome decoding* devised by Schalkwijk *et al.* [401, 402, 403] and modified by Reed and Truong [369, 370, 371] uses a Viterbi-like algorithm to determine the minimum weight error pattern corresponding to the syndrome of the received sequence. We know that the syndrome corresponding to an encoded sequence is zero, and that the syndrome depends only on the channel error or noise sequence. The syndrome decoder finds the minimum Hamming weight error vector that can produce a given syndrome. This error sequence estimate is then used to estimate the transmitted code sequence. Syndrome decoding has been shown to provide some saving in decoder hardware. A summary of the algebraic syndrome decoding proposed by Reed and Truong [369, 370] is given below.

In Chapter 2, an algorithm based on the invariant factor theorem to find the parity-check matrix $\mathbf{H}(D)$ of a rate-$k/n$ encoder, given its generator matrix

$\mathbf{G}(D)$, was outlined. Assume that the matrix $\mathbf{G}(D)$ is basic, i.e., a polynomial matrix with a polynomial right inverse. Given $\mathbf{H}(D)$, the syndrome formation can be expressed as

$$\mathbf{S}(D) = \mathbf{R}(D)\mathbf{H}^T(D), \tag{5.22}$$

and since the syndrome is data-independent, we get

$$\mathbf{S}(D) = \mathbf{E}(D)\mathbf{H}^T(D). \tag{5.23}$$

The decoding task is to find the general solution for $\mathbf{E}(D)$ in (5.23) given that the syndrome is calculated using (5.22).

We can use the invariant factor theorem to write

$$\mathbf{H}^T(D) = \mathbf{L}(D)[\mathbf{I}_{n-k},\ 0]^T\mathbf{T}(D), \tag{5.24}$$

which, when substituted in (5.23), gives

$$\mathbf{S}(D) = \mathbf{E}(D)\mathbf{L}(D)[\mathbf{I}_{n-k},\ 0]^T\mathbf{T}(D). \tag{5.25}$$

Multiplying both sides in (5.25), we get

$$\sigma(D) = \epsilon(D)[\mathbf{I}_{n-k},\ 0]^T, \tag{5.26}$$

where $\sigma(D) = \mathbf{S}(D)\mathbf{T}^{-1}(D)$ and $\epsilon(D) = \mathbf{E}(D)\mathbf{L}(D)$. The component-by-component solution of (5.26) is given by

$$\epsilon_j(D) = \begin{cases} \sigma_j(D) & \text{for } 1 \le \mathrm{j} \le \mathrm{n}-\mathrm{k} \\ t_{j-n+k}(D) & \text{for } \mathrm{n}-\mathrm{k}+1 \le \mathrm{j} \le \mathrm{n}, \end{cases} \tag{5.27}$$

where $t_1(D),\ \ldots,\ t_k(D)$ are arbitrary polynomials in $F[D]$. Finally, the general solution for (5.23) is

$$\begin{aligned} \mathbf{E}(D) &= [e_1(D),\ \ldots,\ e_n(D)] & (5.28) \\ &= [\sigma_1(D),\ \ldots,\ \sigma_{n-k}(D),\ t_1(D),\ \ldots, t_k(D)]\mathbf{L}^{-1}(D). & (5.29) \end{aligned}$$

Reed and Truong [369, 370] have outlined a Viterbi-like recursive decoding algorithm that finds the maximum likelihood estimate $\hat{\mathbf{E}}(D)$ for a truncated $(n,\ k)$ convolutional code over a $q$-ary symmetric channel. This estimate is obtained by minimizing the Hamming weight of the error vector over all possible polynomials $t_1(D), \ldots,\ t_k(D)$. Their algorithm uses a "universal" trellis whose states are the states of the shift register needed to store the delayed versions of the arbitrary function $t(D)$.

## 5.6 VLSI ARCHITECTURES FOR VITERBI DECODING

Ever increasing data rates in digital communication systems have led many a researcher to explore high-speed implementations of the Viterbi algorithm. The phenomenal growth of *very-large-scale-integration* (VLSI) technology has made high-speed single-chip Viterbi decoders a reality. Indeed, many companies are offering such a decoder [367, 65]. The key to a successful high-speed VLSI implementation of the Viterbi algorithm is the ability to extract as much inherent parallelism in the algorithm as possible.

The Viterbi algorithm is essentially sequential in nature and has high computational and storage requirements. However, the type of computations is the same, a desirable feature that results in a highly regular VLSI circuit design. The computation is *add-compare-select (ACS)*, performed at each state at each level. In a *state-serial* implementation, a very high-speed ACS unit is used and the ACS operations corresponding to all the $2^{\nu}$ states must be carried out at each level sequentially before the next received block can be processed to avoid a queue at the decoder input. Thus, the operating speed of the ACS unit and the encoder constraint length $\nu$ determine the maximum data rate achievable by the decoder.

In an alternate approach, a *state-parallel* implementation is obtained by using an ACS unit per state. With $2^{\nu}$ ACS units, a new received block can be processed as soon as the parallel computation is over and the results have been communicated to the appropriate ACS units for the next step. But therein lies the bottleneck of this implementation. The wirings required to connect the ACS units to achieve efficient communication becomes excessive. This introduces another bottleneck when a VLSI implementation of the decoder with $2^{\nu}$ ACS units operating in parallel is desired.

In VLSI design, a trade-off exists between the operating speed and the chip area of the implementation. A judicious choice has to be made based on the desired data rates and the given encoder rate and length. A third approach involves the use of a single ACS unit for a subset of states, thus requiring fewer than $2^{\nu}$ units and striking a balance between the parallelism achieved and the wiring overhead expended.

Recently, many architectures for VLSI implementation of the Viterbi decoder have been proposed. Three levels exist for introducing parallelism, namely, bit-, word-, and algorithm-level, and a detailed survey of solutions for the Viterbi algorithm at all three levels is given by Fettweis and Meyr [143]. For further details, see [167, 168, 255, 214, 166, 73, 282, 438, 65, 140, 142, 141, 334, 295, 390, 372, 462, 17, 59, 84, 107, 418, 419].

## 5.7 EXERCISES

1. Draw the trellis diagram of the (2, 1, 1) encoder with generator sequences $\mathbf{g}^{(1)} = 10$ and $\mathbf{g}^{(2)} = 11$ for an information sequence of length $N = 5$ bits. Find the encoded sequence $\mathbf{v}$ for $\mathbf{u} = 10111$.

2. Use the Viterbi algorithm to find the estimated encoded sequence $\hat{\mathbf{v}}$ corresponding to the encoded sequence $\mathbf{v}$ of Exercise 1 if the error sequence is $\mathbf{e} = 00\ 01\ 00\ 00\ 00\ 00$. Count the number of computations performed.

3. Repeat Exercise 2 for $\mathbf{e} = 00\ 01\ 10\ 00\ 00\ 00$.

4. Repeat Exercise 2 for $\mathbf{e} = 10\ 00\ 00\ 00\ 01\ 00$.

5. Draw the trellis diagram of the (3, 2, 1) encoder with generator sequences $\mathbf{g}_1^{(1)} = 11$, $\mathbf{g}_1^{(2)} = 01$, $\mathbf{g}_1^{(3)} = 11$, $\mathbf{g}_2^{(1)} = 01$, $\mathbf{g}_2^{(2)} = 10$, and $\mathbf{g}_2^{(3)} = 10$ for an information sequence length $N = 5$ blocks (two bits per block). Find the encoded sequence $\mathbf{v}$ for $\mathbf{u} = 10\ 11\ 01\ 10\ 11$.

6. Use the Viterbi algorithm to find the estimated encoded sequence $\hat{\mathbf{v}}$ corresponding to the encoded sequence $\mathbf{v}$ of Exercise 5 if the error sequence is $\mathbf{e} = 000\ 001\ 000\ 000\ 000\ 000$. Count the number of computations performed.

7. The (2, 1, 1) and (3, 2, 1) encoders of Exercises 1 and 5 both have $d_{free} = 3$. Compare the advantages and disadvantages of using either one for single-error correction in terms of the computational complexity.

8. Compute the metric table and the integer metric table for a binary input, 4-ary output DMC with transition probabilities $P(r_i|v_i)$ given in Table 5.2.

9. The encoded sequence in Exercise 1 is transmitted over the binary input, 4-ary output DMC of Exercise 8. The corresponding 4-ary received sequence is

   $$\mathbf{r}_{DMC} = 1_1 1_1\ 0_1 1_2\ 0_2 1_2\ 1_1 0_2\ 1_1 0_1\ 0_1 1_1.$$

   Use the Viterbi algorithm and the integer metric table computed in Exercise 8 to obtain the estimated information sequence.

| | $r_i$ | | | |
|---|---|---|---|---|
| $v_i$ | $0_1$ | $0_2$ | $1_2$ | $1_1$ |
| 0 | 0.5 | 0.4 | 0.09 | 0.01 |
| 1 | 0.01 | 0.09 | 0.4 | 0.5 |

**Table 5.2** Transition probabilities for a binary input, 4-ary output DMC.

10. Compute the approximate bit error probability $P_b(E)$ for the (2, 1, 6) nonsystematic encoders listed in Table A.3 in Section 4.12. Assume that a BSC with transition probability $p = 0.01$ is used. Which encoder has the lowest probability of error?

11. The asymptotic coding gain achievable by an encoder is given by (5.20). Using Tables A.3, A.4, A.5, and A.6 in Section 4.12, find encoders of rates $R = 1/2$, $1/3$, $1/4$, and $2/3$ that achieve an asymptotic coding gain of approximately 3 dB.

12. Write a computer program to implement hard decision Viterbi decoding. Assume that the (2, 1, 2) encoder with generator sequences $\mathbf{g}^{(1)} = 101$ and $\mathbf{g}^{(2)} = 111$ is used. The program should accept as input a received sequence of length $M = 2(N + 2)$ bits, which may be different from a codeword. The output of the program should be the estimated encoded sequence $\hat{\mathbf{v}}$ and the corresponding estimated information sequence $\hat{\mathbf{u}}$.

13. Write a computer program to implement soft decision Viterbi decoding. Assume that the $(2, 1, 2)$ encoder with generator sequences $\mathbf{g}^{(1)} = 101$ and $\mathbf{g}^{(2)} = 111$ is used to encode an information sequence of length $N$ bits, and the codeword is transmitted over the binary input, 4-ary output DMC of Exercise 8. The program should accept as input a 4-ary received sequence of length $M = 2(N + 2)$ symbols, and should produce as output the estimated encoded sequence $\hat{\mathbf{v}}$ and the corresponding estimated information sequence $\hat{\mathbf{u}}$.

# 6

# SEQUENTIAL DECODING

Sequential decoding, proposed by Wozencraft [474], was the first practical technique for decoding convolutional codes. A series of versions of sequential decoding were proposed subsequently: one by Fano [135], called the Fano algorithm, and another independently by Zigangirov [494] and Jelinek [217], called the ZJ or the stack algorithm.

Sequential decoding offers the advantage that its computational complexity is independent of the encoder constraint length. This is in contrast to Viterbi decoding in which the computational complexity increases exponentially with the encoder constraint length. This restricts the use of Viterbi decoding to relatively short constraint length encoders, which in turn makes achieving arbitrarily low error probabilities in practice impossible. Furthermore, the Viterbi algorithm performs a constant number of computations per decoding decision even when only a few or no errors are present in the received sequence. Sequential decoding, on the other hand, adaptively adjusts the number of computations per decoding decision according to the noise level. The disadvantage of sequential decoding, however, is that the number of computations required is a random variable and can be excessively high during heavy noise levels and can even result in a decoding failure or erasure.

## 6.1 THE STACK ALGORITHM

Sequential decoding is essentially based on a tree search mechanism. Hence, the operation of the stack algorithm is best explained using the encoder tree diagram described in Chapter 2. The stack algorithm is outlined below.

**Step 0.** Initialization: load the origin node in the tree in the stack; set the metric to zero.

**Step 1.** Compute the metric of the successors of the path at the top of the stack.

**Step 2.** Delete the path at the top of the stack and insert the new paths in the stack; rearrange the stack in the order of decreasing metric values.

**Step 3.** If the path at the top of the stack ends at a terminal node in the tree, stop. Otherwise, go back to step 1.

A log-likelihood metric, adjusted to account for the different lengths, can be used to determine the best path. The log-likelihood metric for each bit is given by

$$M(r_i \mid v_i) = log_2 \frac{P(r_i \mid v_i)}{P(r\cdot)} - R, \tag{6.1}$$

where $R$ is the code rate. The metric for the first $l$ branches is obtained by adding the bit metrics of all the bits in each of the branches:

$$M([\mathbf{r} \mid \mathbf{v}]_{l-1}) = \sum_{i=0}^{nl-1} log_2 P(r_i \mid v_i) - \sum_{i=0}^{nl-1} log_2 P(r_i) - nlR. \tag{6.2}$$

The first term in the metric is the same as that used in Viterbi decoding. The last two terms are used as adjustment due to the fact that paths of unequal lengths are compared at each stage. This metric was proposed by Fano [135] and is called the *Fano metric*. For a BSC with transition probability $p$, the Fano metric is

$$M(r_i \mid v_i) = \begin{cases} log_2 2p - R & \text{if } \mathrm{r_i} \neq \mathrm{v_i} \\ log_2 2(1-p) - R & \text{if } \mathrm{r_i} = \mathrm{v_i} \end{cases} \tag{6.3}$$

## 6.1.1 An Example Run

Figure 6.1 shows the tree diagram corresponding to the (2, 1, 2) encoder with generator sequences $\mathbf{g}^{(1)} = 101$ and $\mathbf{g}^{(2)} = 111$. The encoder is initially in state 00. Consider the information sequence $\mathbf{u} = 10111$ of length $N = 5$ bits, and the corresponding terminated encoded sequence $\mathbf{v} = 11\ 01\ 00\ 10\ 01\ 10\ 11$, shown highlighted in Figure 6.1.

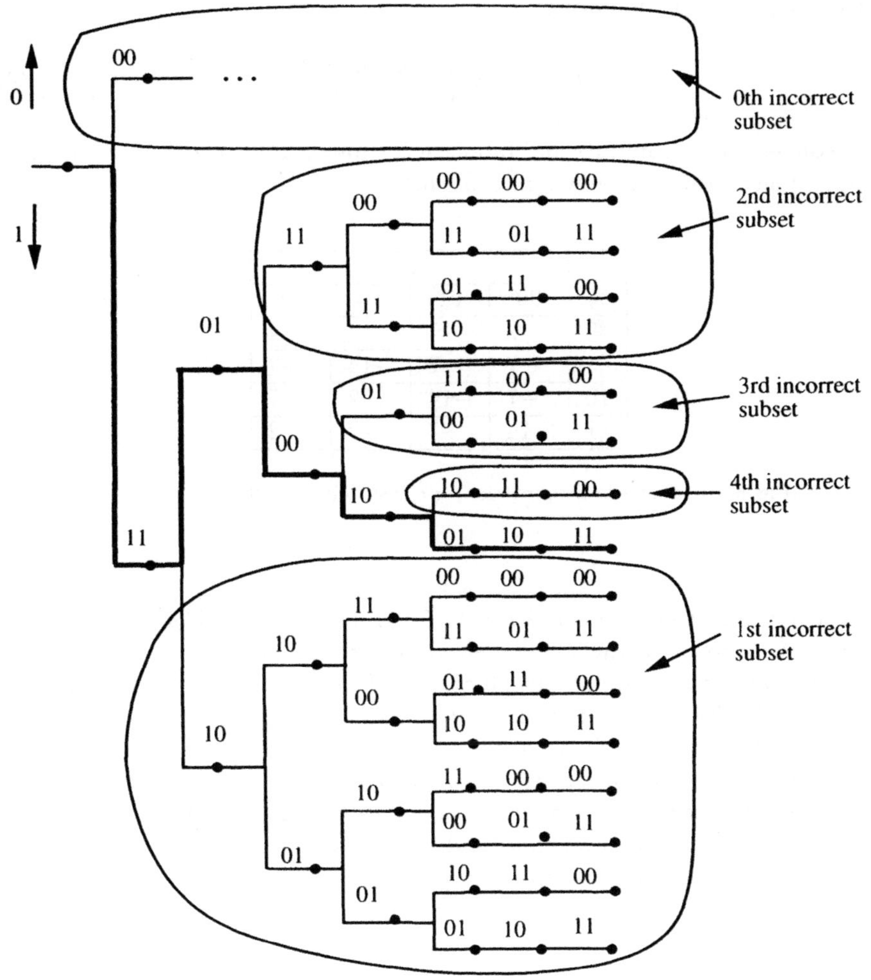


**Figure 6.1** Tree diagram of a (2, 1, 2) encoder.

Assume that a BSC with $p = 0.1$ is being used. The bit metric values are as follows:

$$M(r_i \mid v_i) = \begin{cases} -2.83 & \text{if } r_i \neq v_i \\ 0.35 & \text{if } r_i = v_i \end{cases} \qquad (6.4)$$

The real numbers in (6.4) can be converted to integers by using a scaling factor of 1/0.18 (rounding to the nearest integer may be required). The integer metrics for the BSC are the following:

$$M(r_i \mid v_i) = \begin{cases} -8 & \text{if } \mathrm{r_i} \neq \mathrm{v_i} \\ 1 & \text{if } \mathrm{r_i} = \mathrm{v_i} \end{cases} \tag{6.5}$$

It is convenient to compute and tabulate the metrics for each 2-bit branch. The branch metric table is shown in Table 6.1.

| | $\mathbf{r}_i$ | | | |
|---|---|---|---|---|
| $\mathbf{v}_i$ | 00 | 01 | 10 | 11 |
| 00 | 2 | −7 | −7 | −14 |
| 01 | −7 | 2 | −14 | −7 |
| 10 | −7 | −14 | 2 | −7 |
| 11 | −14 | −7 | −7 | 2 |

**Table 6.1 Integer branch metrics for a BSC with $p = 0.1$.**

Consider the received sequence $\mathbf{r}$ = 11 01 01 11 01 10 11. The stack algorithm begins by comparing the first received branch 11 in step 1 to the first two possible code branches, namely, 00 and 11. From Table 6.1, the metric for the code branch 00 based on the received branch 11 is 2, while that of 11 is −14. The top path, i.e., the origin node in the stack is deleted and the two successor paths are inserted in decreasing order of their metric. At step 2, the path at the top of the stack is expanded to two more paths. The first successor path corresponds to the code branch 01 and has a branch metric of 2, and the second corresponds to the code branch 10 and has a branch metric of −14. The branch metrics are added to the path metric to obtain new path metrics. The path at top of the stack is deleted and the two new successor paths are inserted. There are now three paths on the stack. These are rearranged in decreasing order of their metrics. The step by step progress of the stack algorithm is shown in Table 6.2, where each path is represented by the corresponding information bits with the metrics given in parentheses.

This process is continued for steps 3 through 11. Note that in step 3, there is a tie at the top of the stack. We arbitrarily chose the path labeled 100 for further expansion. Similarly, another tie at step 8 was arbitrarily resolved in favor of the path labeled 1011. These decisions were deliberately made to illustrate the arbitrariness in resolving the tie. In practice, a fixed rule can be used for resolving all the ties, e.g., the path with new information bit 0 can be chosen in all cases. In step 11, the path at top of the stack after rearrangement has metric

| Step 1 | Step 2 | Step 3 | Step 4 | Step 5 | Step 6 |
|---|---|---|---|---|---|
| 1(2) | 10(4) | 100(−3) | 1001(−1) | 10010(1) | 101(−3) |
| 0(−14) | 11(−12) | 101(−3) | 101(−3) | 101(−3) | 100100(−6) |
| | 0(−14) | 11(−12) | 11(−12) | 11(−12) | 11(−12) |
| | | 0(−14) | 0(−14) | 0(−14) | 0(−14) |
| | | | 1000(−17) | 10011(−15) | 10011(−15) |
| | | | | 1000(−17) | 1000(−17) |

| Step 7 | Step 8 | Step 9 | Step 10 | Step 11 |
|---|---|---|---|---|
| 100100(−6) | 1011(−10) | 10111(−8) | 101110(−6) | 1011100(−4) |
| 1010(−10) | 1010(−10) | 1010(−10) | 1010(−10) | 1010(−10) |
| 1011(−10) | 11(−12) | 11(−12) | 11(−12) | 11(−12) |
| 11(−12) | 0(−14 | 0(−14) | 0(−14) | 0(−14) |
| 0(−14) | 10011(−15) | 10011(−15) | 10011(−15) | 10011(−15) |
| 10011(−15) | 1000(−17) | 1000(−17) | 1000(−17) | 1000(−17) |
| 1000(−17) | 1001000(−20) | 1001000(−20) | 1001000(−20) | 1001000(−20) |
| | | 10110(−24) | 10110(−24) | 10110(−24) |

**Table 6.2** Progress of the stack algorithm for a BSC with $p = 0.1$.

equal to −4. It ends in a terminal node, and so we are done. The encoded sequence corresponding to this path, obtained by reading off the output labels on its branches, is $\hat{\mathbf{v}} = 11\ 01\ 00\ 10\ 01\ 10\ 11$, which is the same as $\mathbf{v}$.

## 6.2 ENCODER SELECTION

Optimum performance using sequential decoding is achieved by using encoders with rapidly rising column distance function, i.e., by using ODP encoders. Such encoders ensure fast decoding and minimum erasure probability. See Section 4.12 for listings of ODP encoders.

The error probability is minimized by choosing encoders with large free distance. Since the performance of sequential decoding is independent of the encoder constraint length, encoders with very large constraint length can be chosen. Note that systematic encoders have smaller free distances for the same rate and constraint length than nonsystematic encoders. However, systematic encoders are preferred in some applications since the information sequence is quickly retrieved from the estimated encoded sequence. Sequential decoding allows using systematic encoders with constraint length large enough to achieve the desired value of the free distance without any degradation in computational performance. This is not the case with Viterbi decoding, in which the number of computations is highly sensitive to the encoder constraint length and the error performance is significantly reduced if systematic encoders with smaller free distances are used.

## 6.3 PERFORMANCE ANALYSIS

The main performance measure in sequential decoding is the number of computations performed in decoding. The task of evaluating this number is made more difficult by the possibility of erasures. Hence, the performance analysis of sequential decoding is more difficult than that of Viterbi decoding.

A concept used to characterize the performance of sequential decoding is that of the *jth incorrect subset* of the code tree. It is the set of all nodes branching from the $j$th node on the correct path, $1 \leq j \leq N-1$. For the highlighted correct path, the incorrect subsets are shown in Figure 6.1.

The average number of computations $C_j$ performed in the $j$th incorrect subset has a *Pareto distribution*, given by

$$P[C_j \geq \eta] \approx A\eta^{-\rho}, \quad 0 < \rho < \infty, \quad 0 \leq j \leq N-1, \tag{6.6}$$

where $A$ is a constant depending on the particular sequential decoding algorithm. The parameter $\rho$ is called the *Pareto exponent* and is related to the coding rate $R$ by the following equation:

$$R = \frac{E_0(\rho)}{\rho}, \quad 0 < R < C, \tag{6.7}$$

where $C$ is the channel capacity in bits per use of the channel. The function $E_0(\rho)$ is called the *Gallager function*, and is given by (for a binary input symmetric DMC)

$$E_0(\rho) = \rho - log_2 \frac{1}{2} \sum_j [p(j|0)^{1/1+\rho} + p(j|1)^{1/1+\rho}]^{1+\rho}. \tag{6.8}$$

If $\rho \leq 1$, the average number of computations per decoded branch is unbounded. For a bounded average number of computations, $\rho > 1$, so that

$$R = \frac{E_0(\rho)}{\rho} < E_0(1) = R_0, \tag{6.9}$$

where $R_0$ is called the *computational cut-off rate* of the channel. The code rate $R$ must be less than $R_0$ for the average computational complexity of sequential decoding to be finite. Note that $R_0$ is independent of the encoder constraint length.

The probability of decoded bit error for sequential decoding is the same as that of a maximum-likelihood decoder for high rates ($R > R_0$), and hence is optimum. For low rates ($R < R_0$), the average error probability of sequential decoding is suboptimum. But since $R_0$ is independent of the encoder constraint length, large constraint length encoders can be chosen to obtain a performance level equal to that of Viterbi decoding. For rates near $R_0$, encoders with only a slightly larger constraint length are needed to achieve the same error performance as a Viterbi decoder [285].

# 6.4 OTHER SEQUENTIAL DECODING ALGORITHMS

## 6.4.1 Fano Algorithm

An alternative sequential decoding scheme is the Fano algorithm [135]. The main advantage of this algorithm is that it requires essentially no storage as opposed to the stack algorithm, which needs storage to maintain the stack. Thus, the costly stack rearrangement step is also eliminated. However, the Fano algorithm explores more paths and hence is slower in decoding speed than the stack algorithm.

In the Fano algorithm, the decoder starts with the origin node and examines a sequence of adjacent nodes. At any step, it either moves forward to one of the successor nodes or moves backward to the predecessor of the current node. The branch metric is correspondingly added to or subtracted from the current path metric. The algorithm stops when it reaches a terminal node. Since the algorithm does not explore unconnected nodes as in the stack algorithm, no storage is required to remember the metrics of previously examined paths.

The search through the code tree is guided by a threshold on the path metric. A path is explored in the forward direction as long as its path metric continues to increase. If the metric becomes less than the threshold, the search is backed up and another path is examined. If no paths can be found with a metric above the threshold, the threshold value is lowered and the process is continued. A node in the tree may be visited more than once in the forward direction, with a lower threshold each time to avoid looping in the algorithm. The algorithm eventually reaches a terminal node and stops. For more details on the Fano algorithm, see [135, 285].

### 6.4.2 Generalized Stack Algorithm

Haccoun and Ferguson [172] proposed the *generalized stack algorithm* in which more than one path can be extended at the same time. The algorithm also eliminated redundant paths from the stack by exploiting remergers, as in Viterbi decoding. This bridges the gap between the performance of Viterbi and sequential decoding, and reduces the variability of the computational distribution. However, a price is paid in terms of an increase in the average number of computations and complexity of implementation.

### 6.4.3 Multiple Stack Algorithm

Chevillat and Costello [80] proposed the *multiple stack algorithm* in which erasures are completely eliminated. Thus, its performance can be compared directly with that of Viterbi decoding. In the ordinary stack algorithm, noisy codewords result in many paths with metrics close to each other being extended, i.e., put on the stack. The key feature of the multiple stack algorithm is that the stack size is limited to a fixed number large enough so that only very noisy codewords would cause it to be filled up. The decoding begins as it would in the ordinary stack algorithm, and if it is completed before the stack fills up, the decoded output of the multiple stack algorithm is the same as that of the ordinary one. However, if the stack fills up before a codeword is completely decoded, a predetermined number of the top entries in the stack are transferred to a second, much smaller stack. The decoding proceeds in the second stack using only the transferred entries. If the end of the tree is reached before the second stack fills up, its top path is stored as a tentative decision, the remaining entries in the second stack are eliminated, and decoding resumes in the first stack. If the decoder reaches the end of the tree before the first stack fills up again, the top path in the first stack is compared with the tentative decision reached earlier using the second stack, and the better path is selected as the final output. However, if the first stack fills up again, its top few entries are used to form a new second stack. If the second stack also fills up, its top few entries are transferred to a third stack, and the decoding continues there. Additional stacks are formed until a tentative decoding decision is reached. The decoder always retains the best tentative decision. The remaining paths in the new stack are deleted, and decoding resumes in the previous stack. The decoding terminates if the end of the tree is reached in the first stack, or a computational limit is reached. In the latter case, the best tentative decision is the final output. The multiple stack algorithm has a speed advantage compared to the Viterbi algorithm [299], but has considerable storage and buffer requirements.

### 6.4.4 Sequential Decoding of Punctured Codes

In Chapter 4, punctured codes and encoders were defined. It was shown that a (low) rate-$1/b$ original encoder can be punctured to obtain a (high) rate-$k/n$ punctured encoder. Haccoun and Begin [171] have described the use of the stack algorithm in decoding the punctured code. High-rate punctured codes can be decoded using sequential decoding just as can be done using Viterbi decoding. The decoding is performed on the tree of the original low-rate encoder, using the knowledge of the puncturing matrix $\mathbf{P}$. Decoding proceeds one branch at a time for $k - 1$ intermediate steps involving the original low-rate encoder. At each step, the top node in the stack is extended into its two $b$ symbol branches from which specific symbols are discarded. That is, the metric evaluation for the punctured symbols is inhibited by assigning the same metric increment (usually zero) to them. The likelihood that the decoder is following the correct path is evaluated on a bit-by-bit basis rather than on the basis of $k$ bits like in the usual decoding of a rate-$k/n$ code. This also results in a smaller number of tree nodes being explored. Thus, the stack algorithm can be easily adapted for the decoding of punctured codes, and variable-rate decoding can also be easily implemented.

## 6.5 EXERCISES

1. Draw the tree diagram of the (2, 1, 1) encoder with generator sequences $\mathbf{g}^{(1)} = 10$ and $\mathbf{g}^{(2)} = 11$ for an information sequence of length $N = 5$ bits. Find the encoded sequence $\mathbf{v}$ for $\mathbf{u} = 11101$.

2. Use the stack algorithm on the tree diagram obtained in Exercise 1 to find the estimated encoded sequence $\hat{\mathbf{v}}$ if the error sequence is $\mathbf{e} = 00\ 01\ 00$

00 00 00. Compare the number of decoding steps with the number required by the Viterbi algorithm.

3. Repeat Exercise 2 for $\mathbf{e} = 00\ 01\ 10\ 00\ 00\ 00$.

4. Repeat Exercise 2 for $\mathbf{e} = 10\ 00\ 00\ 00\ 01\ 00$.

5. Draw the tree diagram of the (3, 2, 1) encoder with generator sequences $\mathbf{g}_1^{(1)} = 11$, $\mathbf{g}_1^{(2)} = 01$, $\mathbf{g}_1^{(3)} = 11$, $\mathbf{g}_2^{(1)} = 01$, $\mathbf{g}_2^{(2)} = 10$, and $\mathbf{g}_2^{(3)} = 10$ for an information sequence of length $N = 5$ blocks (two bits per block). Find the encoded sequence $\mathbf{v}$ for $\mathbf{u} = 10\ 11\ 01\ 10\ 11$.

6. Use the stack algorithm on the tree diagram found in Exercise 5 to find the estimated encoded sequence $\hat{\mathbf{v}}$ if the error sequence is $\mathbf{e} = 000\ 001\ 000\ 000\ 000\ 000$. Compare the number of decoding steps with the number required by the Viterbi algorithm.

7. Write a computer program to implement the stack algorithm. Assume that the (2, 1, 2) encoder with generator sequences $\mathbf{g}^{(1)} = 101$ and $\mathbf{g}^{(2)} = 111$ is used. The program should accept as input a received sequence of length $M = 2(N+2)$ bits, which may be different from a codeword. The output of the program should be the estimated encoded sequence $\hat{\mathbf{v}}$ and the corresponding estimated information sequence $\hat{\mathbf{u}}$.

# 7

# MAJORITY-LOGIC DECODING

Majority-logic decoding of convolutional codes, proposed by Massey [308], is a sub-optimum but simple decoding scheme that allows a high-speed implementation. Its performance is sub-optimum since each of its decoding decision is based only on one output constraint length of received bits. This "limited vision" is not there in Viterbi or sequential decoding which, in principle, consider the entire received sequence before making decoding decisions.

## 7.1 SYNDROME GENERATION

In this chapter, we will primarily consider systematic convolutional encoders. For a (2, 1, $m$) systematic encoder, we have $\mathbf{g}^{(1)} = 1\ 0\ 0\ \cdots\ 0$ and hence, $\mathbf{v}^{(1)} = \mathbf{u} * \mathbf{g}^{(1)} = \mathbf{u}$. Therefore, we can call $\mathbf{v}^{(1)}$ the information sequence and $\mathbf{v}^{(2)}$ the parity sequence. The discrete semi-infinite generator matrix of a (2, 1, $m$) systematic encoder is

$$\mathbf{G} = \begin{bmatrix} 1\,g_0^{(2)} & 0\,g_1^{(2)} & 0\,g_2^{(2)} & \cdots & 0\,g_m^{(2)} & & & \\ & 1\,g_0^{(2)} & 0\,g_1^{(2)} & \cdots & 0\,g_{m-1}^{(2)} & 0\,g_m^{(2)} & & \\ & & 1\,g_0^{(2)} & \cdots & 0\,g_{m-2}^{(2)} & 0\,g_{m-1}^{(2)} & 0\,g_m^{(2)} & \\ & & & \ddots & & & & \ddots \end{bmatrix}, \qquad (7.1)$$

and the encoding operation is $\mathbf{v} = \mathbf{uG}$. Since $\mathbf{v}^{(1)} = \mathbf{u}$, the composite encoded sequence $\mathbf{v} = (v_0^{(1)}, v_0^{(2)}, v_1^{(1)}, v_1^{(2)}, \ldots) = (u_0, v_0^{(2)}, u_1, v_1^{(2)}, \ldots)$, in which the information sequence is clearly distinguishable.

**Example 7.1** The generator matrix of the (2, 1, 6) systematic encoder with generator sequences $\mathbf{g}^{(1)} = 1000000$ and $\mathbf{g}^{(2)} = 1010011$ is

$$\mathbf{G} = \begin{bmatrix} 11 & 00 & 01 & 00 & 00 & 01 & 01 & & & \\ & 11 & 00 & 01 & 00 & 00 & 01 & 01 & & \\ & & 11 & 00 & 01 & 00 & 00 & 01 & 01 & \\ & & & & \ddots & & & & & \ddots \end{bmatrix}. \quad (7.2)$$

The information sequence $\mathbf{u} = 1011101$ is encoded using $\mathbf{G}$ to get the composite encoded sequence $\mathbf{v} = 11\ 00\ 10\ 11\ 10\ 00\ 11\ 01\ 01\ 00\ 01\ 01\ 01$. The two encoded sequences are $\mathbf{v}^{(1)} = 1\,0\,1\,1\,1\,0\,1\,0\,0\,0\,0\,0\,0$, and $\mathbf{v}^{(2)} = 1\,0\,0\,1\,0\,0\,1\,1\,1\,0\,1\,1\,1$. Note that $\mathbf{v}^{(1)} = \mathbf{u}$, with terminating zeros.

The parity-check matrix for a (2, 1, $m$) systematic encoder is

$$\mathbf{H} = \begin{bmatrix} g_0^{(2)}\ 1 & & & & & \\ g_1^{(2)}\ 0 & g_0^{(2)}\ 1 & & & & \\ g_2^{(2)}\ 0 & g_1^{(2)}\ 0 & g_0^{(2)}\ 1 & & & \\ \vdots & \vdots & \vdots & & & \\ g_m^{(2)}\ 0 & g_{m-1}^{(2)}\ 0 & g_{m-2}^{(2)}\ 0 & \ddots & & \\ & g_m^{(2)}\ 0 & g_{m-1}^{(2)}\ 0 & & & \\ & & g_m^{(2)}\ 0 & g_{m-1}^{(2)}\ 0 & \cdots & g_0^{(2)}\ 1 \\ & & & \ddots & & \ddots \end{bmatrix}. \quad (7.3)$$

If the codeword $\mathbf{v}$ is transmitted over a BSC and is received as the sequence $\mathbf{r} = (r_0^{(1)}, r_0^{(2)}, r_1^{(1)}, r_1^{(2)}, \ldots)$, we have $\mathbf{r} = \mathbf{v} + \mathbf{e}$, where $\mathbf{e} = (e_0^{(1)}, e_0^{(2)}, e_1^{(1)}, e_1^{(2)}, \ldots)$ is the (composite) error sequence with $e_i^{(j)} = 1$ whenever $r_i^{(j)} \neq v_i^{(j)}$. Recall from Chapter 2 that the syndrome sequence $\mathbf{s} = (s_0, s_1, s_2, \ldots)$ is computed as $\mathbf{s} = \mathbf{r}\mathbf{H}^T$. Hence, we have $\mathbf{s} = \mathbf{v}\mathbf{H}^T + \mathbf{e}\mathbf{H}^T = \mathbf{e}\mathbf{H}^T$, since $\mathbf{G}\mathbf{H}^T = \mathbf{0}$, where $\mathbf{H}^T$ is the syndrome-former.

**Example 7.2** The syndrome-former of the (2, 1, 6) encoder in Example 7.1 is

$$\mathbf{H}^T = \begin{bmatrix} 1 & 0 & 1 & 0 & 0 & 1 & 1 & & \\ 1 & 0 & 0 & 0 & 0 & 0 & 0 & & \\ & 1 & 0 & 1 & 0 & 0 & 1 & 1 & \\ & 1 & 0 & 0 & 0 & 0 & 0 & 0 & \\ & & & \ddots & & & & & \ddots \end{bmatrix}. \quad (7.4)$$

Let the received sequence be $\mathbf{r} = 11\ 00\ 11\ 01\ 10\ 00\ 11\ 01\ 01\ 00\ 01\ 01\ 01$. Then, the syndrome sequence is $\mathbf{s} = 0\ 0\ 1\ 1\ 0\ 1\ 0\ 0\ 1\ 1\ 0\ 0\ 0$.

The received sequence $\mathbf{r}$ consists of the received information sequence $\mathbf{r}^{(1)} = \mathbf{v}^{(1)} + \mathbf{e}^{(1)} = \mathbf{u} + \mathbf{e}^{(1)}$ and the received parity sequence $\mathbf{r}^{(2)} = \mathbf{v}^{(2)} + \mathbf{e}^{(2)}$. Thus, $\mathbf{e}^{(1)}$ is the information error sequence, and $\mathbf{e}^{(2)}$ is the parity error sequence.

In the polynomial notation, a systematic encoder has $\mathbf{g}^{(1)}(D) = 1$, and the encoding operation can be specified as

$$\begin{aligned} \mathbf{v}^{(1)}(D) &= \mathbf{u}(D)\mathbf{g}^{(1)}(D) = \mathbf{u}(D) \\ \mathbf{v}^{(2)}(D) &= \mathbf{u}(D)\mathbf{g}^{(2)}(D). \end{aligned}$$

The information and parity received polynomials, after transmission over a BSC are

$$\begin{aligned} \mathbf{r}^{(1)}(D) &= \mathbf{v}^{(1)}(D) + \mathbf{e}^{(1)}(D) = \mathbf{u}(D) + \mathbf{e}^{(1)}(D) \\ \mathbf{r}^{(2)}(D) &= \mathbf{v}^{(2)}(D) + \mathbf{e}^{(2)}(D). \end{aligned} \tag{7.5}$$

The syndrome polynomial is formed as follows:

$$\mathbf{s}(D) = \mathbf{r}^{(1)}(D)\mathbf{g}^{(2)}(D) + \mathbf{r}^{(2)}(D). \tag{7.6}$$

That is, the received information polynomial is encoded using the second generator polynomial and the received parity polynomial is subtracted from the parity generated using the received information polynomial. Note that in $GF(2)$, subtraction is the same as addition. Substituting from (7.5) in (7.6), we get

$$\begin{aligned} \mathbf{s}(D) &= (\mathbf{u}(D) + \mathbf{e}^{(1)}(D))\mathbf{g}^{(2)}(D) + \mathbf{v}^{(2)}(D) + \mathbf{e}^{(2)}(D) \\ &= \mathbf{u}(D)\mathbf{g}^{(2)}(D) + \mathbf{v}^{(2)}(D) + \mathbf{e}^{(1)}(D)\mathbf{g}^{(2)}(D) + \mathbf{e}^{(2)}(D) \\ &= \mathbf{v}^{(2)}(D) + \mathbf{v}^{(2)}(D) + \mathbf{e}^{(1)}(D)\mathbf{g}^{(2)}(D) + \mathbf{e}^{(2)}(D) \end{aligned}$$

and therefore, we get

$$\mathbf{s}(D) = \mathbf{e}^{(1)}(D)\mathbf{g}^{(2)}(D) + \mathbf{e}^{(2)}(D). \tag{7.7}$$

The syndrome sequence is used in formulating the majority-logic decoding algorithm, as decribed in the next section.

## 7.2 MAJORITY-LOGIC DECODING ALGORITHM

Majority-logic decoding uses an algebraic algorithm that calculates a finite length syndrome based on one constraint length of the received sequence. The syndrome is used to form a set of $J$ orthogonal checksums for each error bit. A syndrome bit or a sum of syndrome bits can be thought of as a *parity-check sum* since it represents a known sum of channel error bits. The parity-check sums are zero if the received sequence is a codeword. A nonzero parity-check sum corresponds to a received sequence with errors.

An error bit $e_j$ is said to be *checked* by a check sum if $e_j$ is included in the sum. A set of $J$ check sums is *orthogonal* on $e_j$ if each check sum checks $e_j$, but no other error bit is checked by more than one check sum. Let $t_{ML} = \lfloor J/2 \rfloor$. Then, the following *majority-logic decoding rule* can be used to estimate the error bit $e_j$: choose the estimate $\hat{e}_j = 1$ if and only if more than $t_{ML}$ of the $J$ check sums orthogonal on $e_j$ have a value 1. If the error bits checked by the $J$ orthogonal check sums contain $t_{ML}$ or fewer errors, the majority-logic decoding rule correctly estimates $e_j$. Consequently, $t_{ML}$ is called the *majority-logic error-correcting capability* of the encoder. The total number of channel error bits checked by the orthogonal check sum equations is called the *effective constraint length* $n_E$ of the encoder.

**Example 7.3** Consider the (2, 1, 6) systematic encoder with generator sequences $\mathbf{g}^{(1)} = 1000000$ and $\mathbf{g}^{(2)} = 1010011$. The first constraint length of syndrome bits are formed by

$$[\mathbf{s}]_6 = [\mathbf{e}]_6 [H]_6^T. \tag{7.8}$$

We can rewrite these after separating the two error streams as

$$[\mathbf{s}]_6^T = \begin{bmatrix} s_0 \\ s_1 \\ s_2 \\ s_3 \\ s_4 \\ s_5 \\ s_6 \end{bmatrix} = \begin{bmatrix} 1 & & & & & & \\ 0 & 1 & & & & & \\ 1 & 0 & 1 & & & & \\ 0 & 1 & 0 & 1 & & & \\ 0 & 0 & 1 & 0 & 1 & & \\ 1 & 0 & 0 & 1 & 0 & 1 & \\ 1 & 1 & 0 & 0 & 1 & 0 & 1 \end{bmatrix} \begin{bmatrix} e_0^{(1)} \\ e_1^{(1)} \\ e_2^{(1)} \\ e_3^{(1)} \\ e_4^{(1)} \\ e_5^{(1)} \\ e_6^{(1)} \end{bmatrix} + \begin{bmatrix} e_0^{(2)} \\ e_1^{(2)} \\ e_2^{(2)} \\ e_3^{(2)} \\ e_4^{(2)} \\ e_5^{(2)} \\ e_6^{(2)} \end{bmatrix}. \tag{7.9}$$

The $7 \times 7$ matrix on the right hand side is called the *parity triangle* and is obtained from the second generator sequence $\mathbf{g}^{(2)}$. The four check sums are $s_0$, $s_2$, $s_5$, and $s_6$. If we write these check sums out explicitly, we get

$$\begin{aligned} s_0 &= e_0^{(1)} & & & & & & & +e_0^{(2)} \\ s_2 &= e_0^{(1)} & & +e_2^{(1)} & & & & & +e_2^{(2)} \\ s_5 &= e_0^{(1)} & & & +e_3^{(1)} & & +e_5^{(1)} & & +e_5^{(2)} \\ s_6 &= e_0^{(1)} & +e_1^{(1)} & & & +e_4^{(1)} & & +e_6^{(1)} & +e_6^{(2)} \end{aligned} . \tag{7.10}$$

Notice that each of the check sum includes $e_0^{(1)}$ and no other error bit is included in more than one check sum. Thus, these four check sums are orthogonal on $e_0^{(1)}$ and can be used with the majority-logic decoding. There are 11 error bits involved in the check sums. Hence, the effective constraint length is $n_E = 11$. Note that the output constraint length $n_A = 14$ for this encoder, which is more than the effective constraint length. Usually, $n_E < n_A$. Here $J = 4$ and $t_{ML} = 2$, so the decoder will correct properly if two or less errors occur in one effective constraint length of the received bits.

Note that each of the check sums in (7.10) is a single syndrome bit. If each check sum bit is a single syndrome bit, and not a sum of syndrome bits, the corresponding encoder is called a *self-orthogonal encoder*. It is easy to check whether an encoder is self-orthogonal or not from the parity triangle in (7.9). Only those syndrome bits can be used as check sums on $e_0^{(1)}$ that have a 1 in the first column of the parity triangle. Thus, $s_1$, $s_3$, and $s_4$ are excluded from consideration. Now, consider the rows in the parity triangle that have a 1 in the first column. Each of the second through the seventh column has exactly one 1 and rest all 0's. This means that no other error bit can occur in more than one check sum. Not all systematic encoders are self-orthogonal, as illustrated by the following example.

**Example 7.4** The parity triangle for the (2, 1, 5) systematic encoder with generator sequences $\mathbf{g}^{(1)} = 100000$ and $\mathbf{g}^{(2)} = 100111$ is

$$\begin{bmatrix} 1 & & & & & \\ 0 & 1 & & & & \\ 0 & 0 & 1 & & & \\ 1 & 0 & 0 & 1 & & \\ 1 & 1 & 0 & 0 & 1 & \\ 1 & 1 & 1 & 0 & 0 & 1 \end{bmatrix}. \tag{7.11}$$

Consider the first, fourth, fifth and the sixth rows. Both fifth and sixth rows have 1's in the second column. Thus, the encoder is not self-orthogonal.

The encoder in Example 7.4 would appear to have only three check sums in a set orthogonal on $e_0^{(1)}$ if only single syndrome bits are considered. However, by considering sums of syndrome bits, the number of check sums in the set orthogonal on $e_0^{(1)}$ can be increased to 4. The four check sums are $s_0$, $s_1 + s_4$, $s_3$, and $s_5$.

The question now is if the set of check sums orthogonal on $e_0^{(1)}$ can be used to estimate other error bits. The answer turns out to be affirmative. A majority-logic decoder starts out by estimating the first error bit. The error bit estimate $\hat{e}_0^{(1)}$ is then subtrated from each syndrome equation it affects to form a *modified syndrome* set $s'_0, s'_1, \ldots, s'_\nu$. Thus, $s'_0 = s_0 + \hat{e}_0^{(1)}$, $s'_1 = s_1 + \hat{e}_0^{(1)}$, etc. In fact, $s_0$ does not need to be modified since it is not used in forming the next set of check sums. The modified syndrome bits $s'_1, \ldots, s'_\nu$ along with the newly calculated syndrome bit $s_{\nu+1}$ are then used to estimate the next error bit $e_1^{(1)}$. In general, the $i$th information error bits $e_i^{(1)}$ is estimated using the following check sums:

$$\begin{array}{rlllllll}
s'_i = & e_i^{(1)} & & & & & & +e_0^{(2)} \\
s'_{i+2} = & e_i^{(1)} & & +e_{i+2}^{(1)} & & & & +e_{i+2}^{(2)} \\
s'_{i+5} = & e_i^{(1)} & & & +e_{i+3}^{(1)} & & +e_{i+5}^{(1)} & +e_{i+5}^{(2)} \\
s_{i+6} = & e_i^{(1)} & +e_{i+1}^{(1)} & & & +e_{i+4}^{(1)} & +e_{i+6}^{(1)} & +e_{i+6}^{(2)}
\end{array} \qquad (7.12)$$

Here we are assume that all the previous estimates are correct. The decoding continues in this manner until all the information error bits are estimated.

We now give the majority-logic decoding algorithm:

**Step 1.** The $(m+1)(n-k)$ syndrome bits in the first constraint length are calculated.

**Step 2.** A set of $J$ orthogonal check sums is formed on each of the $k$ information error bits from the syndrome bits calculated above.

**Step 3.** Each set of $J$ check sums is fed into a majority-logic gate which produces an output of 1 if and only if more than half of its inputs are 1's. The output of each gate is added to the corresponding received bit. This operation inverts those received bits that correspond to an error bit estimate of 1. The output of each gate is also fed back and subtracted from each syndrome it affects.

**Step 4.** The estimated information bit $\hat{u}_0^{(i)} = \hat{r}_0^{(i)} + \hat{e}_0^{(i)}$, $i = 1, \ldots, ,k$ are shifted out of the decoder. The syndrome registers are shifted once to the right, the next block of $n$ received bits is shifted into the decoder, and the next set of $n-k$ syndrome bits is calculated and shifted into the left most stages of the $n-k$ syndrome registers.

**Step 5.** The syndrome registers now contain the modified syndrome bits together with the new set of syndrome bits. The decoder repeats steps 2,3, and 4 and estimates the next block of error bits. All successive blocks of error bits are estimated in the same way.

Since the error estimate is fed back to the decoder to obtain the modified syndrome set, this decoder is known as a *feedback decoder*. Each error bit affects a limited number of check sums. In fact, this number is equal to the constraint length of the code. Hence, the majority-logic decoder calculates syndrome bits in sets of one constraint length only. This set of check sums yields a majority-logic estimate of the error bit on which it is orthogonal, and that error bit is never looked at again. This makes the decoding simple but introduces a number of problems.

First, the majority-logic decoder works correctly only if there are less than $t_{ML}$ errors in the error bits checked by the set of one constraint length syndrome bits. Second, if the decoder makes a mistake in estimating an error bit, this mistake is fed back into the calculation of the next set of syndrome bits, and hence the estimate of the next error bit. This can cause *error propagation* which would not occur if there is no feedback. See Chapter 8 for a description of feedback decoding. Error propagation is discussed in [67].

Even in the absence of feedback, where error propagation has been completely removed, the performance of the majority-logic decoder is less than optimal. This is due to the fact that a final decision on the value of an error bit is made by looking at only one constraint length of syndrome bits. This procedure works fine for single-bit errors. But if the received sequence has more than one error in a constraint length such that their effects overlap, then it is not possible to decode the sequence without errors.

## 7.3 ENCODER SELECTION

Systematic encoders have many desirable properties that make them attractive for use in majority-logic decoding. Since the information sequence appears in tact as part of the encoded sequence, only the information part of the received sequence needs to be corrected and an inversion step is not required. Nonsystematic encoders do not have a distance advantage since they cannot achieve larger minimum distances than systematic encoders. Systematic encoders are easily orthogonalizable (as opposed to nonsystematic encoders), which can be easily tested. Systematic OMD encoders are tabulated in Section 4.12. Massey [308, 309] has identified three main classes of systematic encoders that can be used in majority-logic decoding: uniform (described in Chapter 4), self-orthogonal, and orthogonalizable (trial and error) encoders.

### 7.3.1 Self-Orthogonal Encoders

Massey [308] was the first to construct self-orthogonal encoders.

**Definition 7.1 (Massey)** *A self-orthogonal convolutional encoder is defined to be an encoder in which the set of all parity checks that check an information error bit in the first block is itself a set of parity checks orthogonal on that error bit.*

Robinson and Bernstein [380] devised a more efficient construction based on the notion of difference sets.

### 7.3.2 Orthogonalizable Encoders

Many encoders that are not self-orthogonal are nonetheless orthogonalizable by forming check sums from sums of syndrome bits. Massey [308] called these orthogonalizable encoders *trial-and-error* encoders.

**Definition 7.2 (Massey)** *A rate-1/n convolutional encoder with minimum distance $d_{min}$ is said to be completely orthogonalizable if $d_{min} - 1$ parity checks orthogonal on $e_0^{(1)}$ can be formed.*

In this case, $J = d_{min} - 1$, and $t_{ML} = t_{min}$. That is, any error pattern that is guaranteed to be correctable by the minimum distance of the encoder is also correctable by majority-logic decoding when the encoder can be completely orthogonalized.

## 7.4 PERFORMANCE ANALYSIS

Assume that data is transmitted over a BSC with transition probability $p$. Further, assume that majority-logic decoder is used with a $(2, 1, m)$ encoder. A majority-logic scheme that uses $J$ orthogonal checksums has maximum error-correcting capability $t_{ML} = \lfloor J/2 \rfloor$. The probability of decoded bit error for majority-logic decoding, after decoding the first information block, is

$$P_{b1}(E) \leq \sum_{i=t_{ML}+1}^{n_E} \binom{n_E}{i} p^i (1-p)^{n_E - i}. \tag{7.13}$$

If no error propagation is assumed, the bit error probability for any block, $P_b(E)$, is equal to $P_{b1}(E)$. In practice, error propagation will occur, so the performance of the decoding scheme will be somewhat inferior. For small values of $p$, the error probability in (7.13) is dominated by the first term. Then, assuming perfect feedback, we have

$$P_b(E) \approx \binom{n_E}{t_{ML}+1} p^{t_{ML}+1} = C_{maj}\ p^{t_{ML}+1}, \tag{7.14}$$

where $C_{maj}$ is a constant dependent on the structure of the encoder, and $n_E$ is the effective constraint length over which the orthogonal checksums are formed. In general, for a rate-$k/n$ systematic encoder, the bit error probability is obtained by dividing the expression in (7.14) by $k$.

## 7.5 EXERCISES

1. Determine whether the (2, 1, 1) encoder with generator sequences $\mathbf{g}^{(1)} = 10$ and $\mathbf{g}^{(2)} = 11$ is self-orthogonal by forming a set of check sums that is orthogonal on $e_0^{(1)}$. What is the feedback error-correcting capability $t_{min}$ and the majority-logic error-correcting capability $t_{ML}$ for this encoder?

2. Repeat Exercise 1 for the (2, 1, 5) encoder with generator sequences $\mathbf{g}^{(1)} = 100000$ and $\mathbf{g}^{(2)} = 110101$.

3. Repeat Exercise 1 for the (2, 1, 6) encoder with generator sequences $\mathbf{g}^{(1)} = 1000000$ and $\mathbf{g}^{(2)} = 1100101$.

4. Find the parity-check matrix $\mathbf{H}$ for the (2, 1, 1) encoder with generator sequences $\mathbf{g}^{(1)} = 10$ and $\mathbf{g}^{(2)} = 11$. Write the equations for the first constraint length of syndrome bits $s_0$, $s_1$.

5. Determine the effective constraint length of the (2, 1, 17) systematic encoder with $\mathbf{g}^{(2)} = 101000\ 010000\ 010011$.

6. Compute the bit error probability $P_b(E)$ of a majority-logic decoder that uses the (2, 1, 5) encoder of Exercise 2 to decode codewords transmitted on a BSC with transition probability $p = 0.01$.

7. Repeat Exercise 6 for the (2, 1, 6) encoder of Exercise 3.

8. Write a computer program to implement majority-logic decoding. Assume that the (2, 1, 6) encoder of Exercise 3 is used. The program should accept as input a received sequence of length $M = 2(N + 2)$ bits, which may be different from a codeword. The output of the program should be the estimated information sequence $\hat{\mathbf{u}}$.

# 8

# TABLE LOOK-UP DECODING

It is desirable to develop a decoding scheme with speed comparable to that of majority-logic decoding, and error-correcting capability approaching that of Viterbi or sequential decoding. For a given encoding rate and constraint length, this requirement can be met by using table look-up decoding. For systematic encoders, it achieves the full feedback error-correcting capability $t_{min}$ without any requirements such as orthogonalizability. Hence, encoders with much smaller constraint lengths can be used. For instance, rate one-half completely orthogonalizable encoders with the same $d_{min}$ have larger $m$ than rate one-half OMD encoders, and rate one-half self-orthogonal encoders have even larger $m$. Table look-up decoding can also be used with nonsystematic encoders, and error-correcting capability $t_j \geq t_{min}$ is achievable if received subsequences larger than one output constraint length are used in each decoding step.

The idea underlying table look-up decoding is the following: a table is used to store correction information precomputed based on known error patterns. A finite length syndrome is used as an address into the table. Many error patterns generate the same syndrome and the one with minimum number of symbols in error is typically stored in the table. The correction process is iterative, with results of the current iteration being fed back before the next iteration begins. A separate inversion step is used to recover the information sequence corresponding to the corrected received sequence.

## 8.1 FEEDBACK DECODING

The syndrome is a data-independent indicator of errors in the received sequence. For a (2, 1, $m$) encoder with generators $\mathbf{g}^{(1)}$ and $\mathbf{g}^{(2)}$, the composite generator sequence $\mathbf{g}$ in (2.7) and the parity-check sequence $\mathbf{h}$ in (2.13) have the property that $\langle \mathbf{g}, \mathbf{h} \rangle = 0$, where $\langle \cdot, \cdot \rangle$ denotes their inner product. That is,

$$\langle \mathbf{g}, \mathbf{h} \rangle = \sum_{i=0}^{m} \left( g_i^{(1)} g_{m-i}^{(2)} + g_i^{(2)} g_{m-i}^{(1)} \right) = 0. \tag{8.1}$$

Further, any $n_A$-bit encoded subsequence $\mathbf{v}_{[i,i+m]}$ satisfies $\langle \mathbf{v}_{[i,i+m]}, \mathbf{h} \rangle = 0$. Therefore, each symbol in the syndrome sequence $\mathbf{s} = (s_0, s_1, s_2, \ldots)$ can be computed as follows:

$$s_i = \langle \mathbf{r}_{[i,i+m]}, \mathbf{h} \rangle. \tag{8.2}$$

Figure 2.3 shows a shift register implementation of this computation. Linearity of convolutional codes allows us to reformulate (8.2) as

$$s_i = \langle \mathbf{e}_{[i,i+m]}, \mathbf{h} \rangle. \tag{8.3}$$

### 8.1.1 Correction Table Generation

Feedback decoding performs correction on a received block $\mathbf{r}_i$ based on a received subsequence $\mathbf{r}_{[i,i+m]}$ of size one output constraint length $n_A$. Recall from Chapter 3 that the "spread" among the encoded subsequence of this size is given by the minimum distance $d_{min}$. That is, a nearest-neighbor decoding scheme would be able to correctly decode a block if there are $t_{min}$ or fewer bits in error in the $n_A$-bit received subsequence $\mathbf{r}_{[i,i+m]}$. It can be implemented using a table that stores correction information. Each error pattern with $t_{min}$ or fewer bits in errors can be distinguished using the syndrome subsequence corresponding to $\mathbf{r}_{[i,i+m]}$. Let this finite length syndrome subsequence $\mathbf{s}_{[i,i+m]}$ be called the *s-address*. All possible $n_A$-bit error patterns $\mathbf{e}_{[i,i+m]}$ with $t_{min}$ or fewer bits in error are used to generate s-addresses as follows:

$$\mathbf{s}_{[i,i+m]} = \mathbf{e}_{[i,i+m]} [\mathbf{H}^T]_m, \tag{8.4}$$

where $[\mathbf{H}^T]_m$ is the syndrome-former $\mathbf{H}^T$ truncated after $m+1$ columns.

**Example 8.1** Consider the (2, 1, 2) encoder with generator sequences $\mathbf{g}^{(1)} = 110$ and $\mathbf{g}^{(2)} = 111$. It has $d_{min} = 3$, and hence, $t_{min} = 1$. That is, it can be used to correct single-bit errors in $n_A = 6$-bit received subsequences. The correction information is generated in the following manner:

$$\begin{bmatrix} 000 \\ 111 \\ 110 \\ 011 \\ 011 \\ 001 \\ 001 \end{bmatrix} = \left[\begin{array}{|cc|cccc} 0 & 0 & 0 & 0 & 0 & 0 \\ 1 & 0 & 0 & 0 & 0 & 0 \\ 0 & 1 & 0 & 0 & 0 & 0 \\ 0 & 0 & 1 & 0 & 0 & 0 \\ 0 & 0 & 0 & 1 & 0 & 0 \\ 0 & 0 & 0 & 0 & 1 & 0 \\ 0 & 0 & 0 & 0 & 0 & 1 \end{array}\right] \begin{bmatrix} 1 & 1 & 1 \\ 1 & 1 & 0 \\ & 1 & 1 \\ & 1 & 1 \\ & & 1 \\ & & 1 \end{bmatrix}.$$

Only the bits between the vertical bars are stored for each s-address. The corresponding correction table is shown in Table 8.1. Table entries marked (–, –) indicate addresses not generated by single-bit errors.

| s-address | $c^{(1)}$ | $c^{(2)}$ | s-address | $c^{(1)}$ | $c^{(2)}$ |
|---|---|---|---|---|---|
| 000 | 0 | 0 | 100 | – | – |
| 001 | 0 | 0 | 101 | – | – |
| 010 | – | – | 110 | 0 | 1 |
| 011 | 0 | 0 | 111 | 1 | 0 |

**Table 8.1** Table for correcting single-bit errors using a (2, 1, 2) encoder and feedback decoding.

It is not necessary to store the entire $n_A$-bit error pattern that generates the s-address. If the correction is performed on the block $\mathbf{r}_i$ in each step, correction information $\mathbf{e}_i$ for only that block needs to be stored. It is denoted by $\mathbf{c} = (c^{(1)}, c^{(2)})$ in Table 8.1.

## 8.1.2 Error Correction Using Table Look-Up

An s-address $\mathbf{s}_{[i,i+m]}$ is computed based on the received subsequence $\mathbf{r}_{[i,i+m]}$, called the *correction window*. Error correction is performed on the received block $\mathbf{r}_i = (r_i^{(1)}, r_i^{(2)})$, called the *correction pair*.

**Example 8.2** Let $\mathbf{v}_{[0,5]} = 111110001001$ be a 12-bit encoded sequence which is received as $\mathbf{r}_{[0,5]} = 111110011001$. The first s-address based on $\mathbf{r}_{[0,2]} = 111110$ is $\mathbf{s}_{[0,2]} = 000$. This implies no correction is necessary. The next s-address, based on $\mathbf{r}_{[1,3]} = 111001$, is $\mathbf{s}_{[1,3]} = 001$. Table look-up using this s-address yields the correction information $\mathbf{c} = 00$, which, when added to the the correction pair $\mathbf{r}_1$, does not change anything. The next s-address, based on $\mathbf{r}_{[2,4]} = 100110$, is $\mathbf{s}_{[2,4]} = 011$. Table look-up using this s-address yields the correction information $\mathbf{c} = 00$, which, when added to the the correction pair $\mathbf{r}_2$, also does not change anything. The next s-address, based on $\mathbf{r}_{[3,5]} = 011001$, is $\mathbf{s}_{[3,5]} = 110$. Table look-up using this s-address yields the correction information $\mathbf{c} = 01$, which, when added to the the correction pair $\mathbf{r}_3 = 01$, changes the received subsequence to $\mathbf{r}_{[3,5]} = 001001$. The s-address is recomputed to be $\mathbf{s}_{[3,5]} = 000$. The next two s-addresses are $\mathbf{s}_{[4,6]} = 000$ and $\mathbf{s}_{[5,7]} = 000$, assuming all-zero received blocks after $\mathbf{r}_5$. Thus, the corrected received sequence, or the estimated encoded sequence is $\hat{\mathbf{v}}_{[0,5]} = 111110001001$, the same as $\mathbf{v}$.

The recalculation of s-address based on the corrected received sequence is the feedback step. All future decoding decisions are based on the new syndrome. Feedback decoding suffers from error propagation which results when error correction is not perfect. A decoding error may result in further errors being introduced into the received sequence, and this may continue until some steps are taken to stop the propagation of errors. One possibility is to not use feedback at all, and the resulting decoding scheme is called *definite decoding* [377]. The syndrome calculation is based on a larger block of the received bits, but a decoding error affects only a finite number of future decoding decisions and error propagation is avoided. Another method to limit error propagation is to select codes with good resynchronization properties so that error propagation is limited and the decoder can resume proper operation once a span of error-free bits are received.

## 8.2 TABLE LOOK-UP DECODING

Feedback decoding as described in the previous section is limited in performance since each decoding decision is based only on one output constraint length of the received subsequence. Recall from Chapter 3 that for a given encoding rate and constraint length, the column distance function increases beyond $d_{min}$ if encoded subsequences longer than $n_A$-bits are considered. In this section, table look-up decoding based on received subsequences larger than $n_A$-bits is described. Many researchers have studied such table look-up decoding schemes [196, 454, 82, 54, 115, 111].

Let each decoding decision be based on the received subsequence $\mathbf{r}_{[i,i+j]}$. The corresponding s-address is $\mathbf{s}_{[i,i+j]}$. All possible $2(j+1)$-bit error patterns $\mathbf{e}_{[i,i+j]}$ with $t_j$ or fewer bits in error are used to generate s-addresses as follows:

$$\mathbf{s}_{[i,i+j]} = \mathbf{e}_{[i,i+j]}[\mathbf{H}^T]_j, \tag{8.5}$$

where $[\mathbf{H}^T]_j$ is the syndrome-former $\mathbf{H}^T$ truncated after $j+1$ columns. Hence, the column distance function $d_j$ and the corresponding error-correcting capability $t_j$ govern the performance of this scheme.

Correction information is precomputed using error patterns with $t_j$ or fewer bits in errors, and is stored in a correction table. We first consider single-bit errors, followed by two-bit errors, etc. The table generation continues for higher order errors until all the entries in the table are filled. Note that the correction table size in this case is $2^{j+1}$ entries, with two bits per entry. The table look-up decoding algorithm is as follows:

**Step 0.** Initialize $\mathbf{s}_{[i,i+j-1]} = \mathbf{0}$.

**Step 1.** Shift in two new received bits $\mathbf{r}_{i+j} = (r^{(1)}_{i+j}, r^{(2)}_{i+j})$.

**Step 2.** Calculate the syndrome bit $s_{i+j}$ based on $\mathbf{r}_{[i+j-\nu,i+j]}$. If $\mathbf{s}_{[i,i+j]} = \mathbf{0}$ go to step 1, else do a table look-up to obtain $\mathbf{c} = (c^{(1)}, c^{(2)})$.

**Step 3.** Perform the correction $\mathbf{r}_i = \mathbf{r}_i + \mathbf{c}$.

**Step 4.** Recompute the s-address $\mathbf{s}_{[i,i+j]}$. Go to step 1.

The estimated encoded sequence $\hat{\mathbf{v}}$ is the final corrected received sequence $\mathbf{r}$. The error correction is separate from inversion. The inversion step to obtain the estimated information sequence $\hat{\mathbf{u}}$ depends on the type of encoder being used. A systematic encoder does not require any inversion. In fact, for systematic encoders, only one bit $c^{(1)}$ of correction information needs to be stored for each s-address, thus reducing the table size. Other encoders such as the QLI, ELI, and nonsystematic locally invertible encoders do require inversion, but are relatively easy to invert. See Chapter 4 for details.

## 8.2.1 Practical Considerations

The implementation of table look-up decoding requires attention to some practical issues. Typically, a correction table generated using $t_j$ or fewer bit error patters is not completely filled. That is, there are s-addresses that are not generated by any error pattern with $t_j$ or fewer bits in error. This allows us to "fill up" the table with correction information corresponding to many error patterns with more than $t_j$ bits in error. Consider the correction table for the (2, 1, 2) encoder with generator sequences $\mathbf{g}^{(1)} = 101$ and $\mathbf{g}^{(2)} = 111$, shown in Tabel 8.2. This encoder has $d_{free} = d_5 = 5$. Hence, a 6-bit s-address is used to correct all up to $t_{free} = t_5 = $ 2-bit errors. All nonzero entries in the columns marked **c** correspond to a correction required in one or both bits of the least recent block $\mathbf{r}_i$ in the correction window $\mathbf{r}_{[i,i+5]}$.

| $\mathbf{s}_{[i,i+5]}$ | **c** | $\mathbf{s}_{[i,i+5]}$ | **c** | $\mathbf{s}_{[i,i+5]}$ | **c** | $\mathbf{s}_{[i,i+5]}$ | **c** |
|---|---|---|---|---|---|---|---|
| 000000 | 00 | 010000 | 11 | 100000 | | 110000 | |
| 000001 | 00 | 010001 | 00 | 100001 | | 110001 | |
| 000010 | 00 | 010010 | 00 | 100010 | 01 | 110010 | 10 |
| 000011 | 00 | 010011 | 00 | 100011 | | 110011 | |
| 000100 | 00 | 010100 | 00 | 100100 | 10 | 110100 | 01 |
| 000101 | 00 | 010101 | 00 | 100101 | | 110101 | |
| 000110 | 00 | 010110 | 00 | 100110 | 01 | 110110 | 10 |
| 000111 | 00 | 010111 | 00 | 100111 | | 110111 | |
| 001000 | 00 | 011000 | | 101000 | 01 | 111000 | 10 |
| 001001 | 00 | 011001 | 00 | 101001 | 01 | 111001 | 10 |
| 001010 | 00 | 011010 | 00 | 101010 | 01 | 111010 | 10 |
| 001011 | 00 | 011011 | 00 | 101011 | 01 | 111011 | 10 |
| 001100 | | 011100 | 00 | 101100 | 10 | 111100 | 01 |
| 001101 | 00 | 011101 | 00 | 101101 | 01 | 111101 | 10 |
| 001110 | 00 | 011110 | 00 | 101110 | | 111110 | |
| 001111 | 00 | 011111 | 00 | 101111 | 01 | 111111 | 10 |

**Table 8.2** Correction table for a (2, 1, 2) encoder based on 6-bit s-address.

In Table 8.2, the all-zero entries in the columns marked **c** are generated by single- or double-bit errors. The entries left blank correspond to s-addresses not generated by any single- or double-bit errors. It is possible to improve the performance of table look-up decoding based on this table by considering 3-bit error patterns as long as any nonzero correction information either is the same as the nonzero entry already present in the table or falls on the blank entries.

For example, the 3-bit error $\mathbf{e}_{[i,i+5]} = 111000\ 000000$ generates the s-address $\mathbf{s}_{[i,i+5]} = 001100$, which is currently blank. Hence, we can store $\mathbf{c} = 11$ at that s-address, and table look-up decoding would perform a proper correction whenever this 3-bit error pattern occurs. On the other hand, another 3-bit error $\mathbf{e}_{[i,i+5]} = 110100\ 000000$ generates the s-address $\mathbf{s}_{[i,i+5]} = 000100$, which already has an entry $\mathbf{c} = 00$, corresponding to the 2-bit error $\mathbf{e}_{[i,i+5]} = 000011\ 000000$. Two or more error patterns that generate the same s-address but require different correction information are said to be in *conflict*. The correction information corresponding to 3-bit error patterns is stored in the table only if there is no conflict.

## 8.3 ENCODER SELECTION

For a given correction table size of $2^{j+1}$ entries (s-address length of $j+1$-bits), we wish to select an encoder with largest column distance function $d_j$ of order $j$. Such an encoder would maximize the error-correcting capability $t_j$ for a given table size. It has been observed that OFD or ODP encoders do not always meet this criterion. For example, an error-correcting capability $t_{15} = 4$ is achieved by the (2, 1, 7) encoder with generator sequences $\mathbf{g}^{(1)} = 10101001$ and $\mathbf{g}^{(2)} = 11101100$ at order $j = 15$. However, this encoder is not ODP, since it has $d_7 = 5$, whereas the (2, 1, 7) ODP encoder in Table A.3 of Section 4.12 has $d_7 = 6$. However, the ODP encoder has $t_j = 4$ for some $j > 15$ and requires a larger correction table for achieving the same error-correcting capability.

## 8.4 PERFORMANCE ANALYSIS

Assume that codewords are transmitted over a BSC with transition probability $p$. Further, assume that an optimum feedback decoder is used with a (2, 1, $m$) encoder. Then $t_{min} = \lfloor (d_{min} - 1)/2 \rfloor$ is the maximum number of errors that are guaranteed to be corrected. The encoded bit error probability after correction of the first block of two encoded bits is

$$P_{b1}(E) \leq \frac{1}{2} \sum_{i=t_{min}+1}^{n_A} \binom{n_A}{i} p^i (1-p)^{n_A - i}. \tag{8.6}$$

If no error propagation is assumed, the bit error probability for any block, $P_b(E)$, is equal to $P_{b1}(E)$. In practice, error propagation will occur, so the performance of the decoding scheme will be somewhat inferior. For small values of $p$, the error probability in equation (8.6) will be dominated by the first term.

Then, assuming perfect feedback, we have

$$P_b(E) \approx \frac{1}{2} \binom{n_A}{t_{min}+1} p^{t_{min}+1} = C_{fbk}\ p^{t_{min}+1}, \tag{8.7}$$

where $C_{fbk}$ is a constant dependent on the structure of the encoder.

Now, assume that a table look-up decoder is used with a correction window of length $2(j+1)$ bits. The encoded bit error probability after correction of the first block of two encoded bits is

$$P_{b1}(E) \leq \frac{1}{2} \sum_{i=t_j+1}^{2(j+1)} \binom{L}{i} p^i (1-p)^{2(j+1)-i}. \tag{8.8}$$

Again, if no error propagation is assumed, then the bit error probability for any block, $P_b(E)$, is equal to $P_{b1}(E)$. For small values of $p$, the error probability in (8.8) will be dominated by the first term, and we have

$$P_b(E) \approx \frac{1}{2} \binom{2(j+1)}{t_j+1} p^{t_j+1} = C_t\ p^{t_j+1}, \tag{8.9}$$

where $C_t$ is a constant dependent on the structure of the encoder and the correction table. Note that if $j = m$, then $t_j = t_{min}$, and the error probability in (8.9) is the same as that of optimum feedback decoding, and if $j$ is large enough to allow $t_j = t_{free}$, the error probability is of the same order as that of Viterbi decoding.

For a correction table large enough to guarantee the correction of up to $t_j$-bit errors, it is usually possible to correct many higher order errors as well. Let $\tau$ be the number of errors beyond the guaranteed limit $t_j$ that can be partially corrected by a table, and let $\beta_i$ represent the fraction of higher order error patterns that generate a table conflict and are not corrected. Then, the encoded bit error probability of table look-up decoding is

$$\begin{aligned} P_b(E) \leq\ & \tfrac{1}{2} \sum_{i=t_j+1}^{t_j+\tau} \beta_i \binom{2(j+1)}{i} p^i (1-p)^{2(j+1)-i} \\ +\ & \tfrac{1}{2} \sum_{i=t_j+\tau+1}^{2(j+1)} \binom{2(j+1)}{i} p^i (1-p)^{2(j+1)-i}, \end{aligned} \tag{8.10}$$

where $0 \leq \beta_i < 1$ for $t_j + 1 \leq i \leq t_j + \tau$.

# 8.5 OTHER APPROACHES TO TABLE LOOK-UP DECODING

## 8.5.1 Effects of Soft Decisions

Koorapaty *et al.* [251, 250] have proposed a table look-up decoding scheme that incorporates soft decisions in it. $Q$-level quantization is performed at the receiver. A hard quantized received sequence is also formed. Error correction is carried out on this sequence by computing the syndrome and performing a table look-up. Soft decision outputs and precomputed correction table information can be used to resolve conflicts and choose the most probable correction pattern. The table is modified to store additional information about error patterns that generate conflicts. The table size in this case is dependent on the encoder constraint length as well as the length of the correction window. Thus, use of nonsystematic encoders is desirable.

## 8.5.2 Selective Look-Ups

The table look-up decoding algorithm described above performs a look-up every time the computed s-address is non-zero. This means that the correction table is accessed so long as an error is within the correction window, whether the erroneous bits are or are not in the correction pair. The speed of table look-up decoding is governed by the number and speed of table accesses necessary to correct the entire received sequence. The number of table look-ups can be reduced significantly by observing that the correction information is nonzero almost always when the leading bit of the s-address is a 1. This is the case when both the generator sequences satisfy $g_0^{(1)} = g_0^{(2)} = 1$. The table look-up algorithm can be modified to perform a table look-up only when the leading bit of the s-address is a 1. Thus, by restricting the choice of encoders used with table look-up decoding to those that meet this condition and by modifying the decoding algorithm, a much faster decoding speed is achieved. A price is paid, however, in performance since the modified algorithm can not detect all the errors that the original algorithm can. For example, the s-address $\mathbf{s}_{[i,i+5]} =$ 010000 in Table 8.2 has nonzero correction information, but its leading bit is zero. Hence, the modified decoding would not catch this error.

### 8.5.3 Multiple Look-Ups

The difference between two conflicting error patterns is in the positions of the erroneous bits. These positions are known at the time of correction table generation. Table look-up decoding with multiple look-ups makes use of this fact, and is based on modifying the correction table by adding two or more fields to each entry that indicate the positions of the errors. At the time of performing corrections, the extra information is used for shifting the correction window to make a second or a third table look-up. Thus, multiple look-ups allow correction of conflicting error patterns.

## 8.6 EXERCISES

1. What is the feedback error-correcting capability $t_{min}$ of the (2, 1, 1) encoder with generator sequences $\mathbf{g}^{(1)} = 10$ and $\mathbf{g}^{(2)} = 11$? Make a correction table that can be accessed using a 2-bit s-address and allows correction of error patterns with $t_{min}$ or fewer errors.

2. What is the error-correcting capability $t_j$ of the (2, 1, 4) encoder with generator sequences $\mathbf{g}^{(1)} = 10011$ and $\mathbf{g}^{(2)} = 11011$ for $j = 13$? Make a correction table that can be accessed using a $j + 1 = 14$-bit s-address and allows correction of error patterns with $t_j$ or fewer errors.

3. From Table A.3 in Section 4.12, determine an encoder of smallest memory order that has an error-correcting capability $t_j = 4$. This would require computation of the column distance function of the encoders.

4. Consider the (2, 1, 7) encoder with generator sequences $\mathbf{g}^{(1)} = 10101001$ and $\mathbf{g}^{(2)} = 11101100$. Determine the order $j$ of its column distance function at which $d_j = 9$. How does this value compare with that found in Exercise 3.

5. Write a computer program to implement table look-up decoding. Assume that the (2, 1, 2) encoder with generator sequences $\mathbf{g}^{(1)} = 101$ and $\mathbf{g}^{(2)} = 111$ is used. This encoder has $d_{free} = d_5 = 5$, and can correct one and two bit errors. The table size (i.e., the syndrome length) should be large enough to accommodate all one and two bit errors without conflicts. The program should accept as input a received sequence of length $M = 2(N + 2)$ bits, which may be different from a codeword. The output of the program should be the estimated encoded sequence $\hat{\mathbf{v}}$ and the corresponding estimated information sequence $\hat{\mathbf{u}}$.

# PART IV

# APPLICATIONS

# 9

# FIXED-REDUNDANCY ERROR CONTROL SCHEMES

Channel coding is implemented in a variety of ways in a digital communication system. A simple approach, known as *automatic-repeat-request* (ARQ), is to perform error detection at the receiver, and request retransmission of the packet if errors are detected [35, 36, 63, 286]. Conventional ARQ schemes require a feedback channel to send acknowledgements (positive or negative). Under good channel conditions, the throughput of ARQ schemes is very high, limited only by the rate of the error-detecting code. However, as the channel noise increases, the throughput degrades rapidly as a result of an increase in retransmission frequency. In the absence of a feedback channel, error control has to rely on *forward error correction* (FEC) alone. Various convolutional coding and decoding techniques discussed in this book so far are FEC schemes. FEC schemes offer a constant throughput, equal to the coding rate, but their reliability degrades as the channel noise increases.

If a feedback channel is available, the performance of an ARQ scheme can be improved by embedding FEC capability in it. This technique results in a ***hybrid-ARQ*** scheme. A common example is type-I hybrid-ARQ [291]. The advantage of using FEC is that many errors are corrected, and retransmissions are requested only if uncorrectable errors are detected. This results in a reduction in the overall retransmission frequency and hence improves throughput during noisy channel conditions. Both conventional ARQ and type-I hybrid-ARQ are ***fixed-redundancy*** error control schemes. These schemes are suitable for channels with fairly constant average noise levels, i.e., for the so-called stationary channels. In this part of the book, an encoded sequence is equivalent to a packet or a message.

# 9.1 CONVENTIONAL ARQ SCHEMES

Conventional ARQ schemes are mainly of two types: stop-and-wait, and continuous. The latter is also of two types: go-back-N ARQ, and selective-repeat ARQ. These schemes are briefly described in this section, and their reliability and throughput are qualitatively compared. Detailed work on these schemes can be found in the following references: [430], [307], [136], [393].

## 9.1.1 Stop-and-wait ARQ

This is the simplest form of ARQ protocol, in which the transmitter is idle after sending a packet until it gets an acknowledgement from the receiver. The packet is stored in the transmitter buffer pending the receiver's response. The receiver sends a positive acknowledgement (ACK) if no errors are detected in the packet; otherwise, it sends a negative acknowledgement (NAK). The transmitter discards the packet stored in its buffer and sends out the next packet if it gets an ACK. If, however, a NAK is received, the packet which is stored in the buffer is retransmitted. This process is repeated until the packet is successfully received and acknowledged by the receiver.

This strategy is obviously inefficient, especially when the round-trip delay is significantly large, as is the case, for example, in satellite channels. It is not without virtue, however. The instrumentation complexity required to implement stop-and-wait ARQ is minimal. No buffer space is needed at the receiver. The only hardware necessary is that needed for error detection.

Assume that an $(n,\ k)$ block code is being used. Let $P_c$ denote the probability that the received packet contains no error. Further, let $P_d$ denote the probability that the packet contains a detectable error pattern. And finally, let $P_e$ denote the probability that the received packet contains undetectable error pattern. Note that $P_c + P_d + P_e = 1$. A received packet will be accepted by the receiver only if it contains no error or if it contains an undetectable error. Hence, the probability of acceptance is $P = P_c + P_e$. Then, the probability $P(E)$ that the receiver commits an error is given by

$$P(E) = \frac{P_e}{P_c + P_e}. \tag{9.1}$$

The throughput of an ARQ protocol is defined as the ratio of the average number of information bits successfully accepted by the receiver per unit time to the total number of bits that could be transmitted per unit time. Let the time that the transmitter is idle while waiting for an acknowledgement from the receiver be denoted by $T$, and let the transmission rate in bits per second be denoted by $\tau$. Then, the total number of bits that the transmitter can send in one round-trip delay time is $n + T \cdot \tau$. Taking the retransmissions into account, the average number of bits that the transmitter could have transmitted in the time it took the packet to be successfully received is

$$\begin{aligned} T_{SW} &= (n + T \cdot \tau)P + 2(n + T \cdot \tau)P(1-P) \\ &\quad +3(n + T \cdot \tau)P(1-P)^2 + \cdots \\ &= \frac{n + T\tau}{P}. \end{aligned} \tag{9.2}$$

Therefore, the throughput of a stop-and-wait ARQ scheme is

$$\eta_{SW} = \frac{k}{T_{SW}} = \frac{P}{1 + T\tau/n}\left(\frac{k}{n}\right). \tag{9.3}$$

On a BSC with transition probability $p$, the probability of correctly receiving an $n$-bit packet is $P_c = (1-p)^n$. If an $(n, k)$ block code is used as the error-detecting code, the probability of undetected error is overbounded by $P_e \leq 1/2^{(n-k)}$ (see Appendix B).

**Example 9.1** Assume that a (511, 493) block code is used on a BSC with transition probability $p = 0.001$. Then, $P_c \approx 0.6$, and $P_e = 3.8 \times 10^{-6}$. We see that $P_e \ll P_c$. Hence, $P \approx P_c$, and $P(E) = P_e/P = 6.3 \times 10^{-6}$.

### 9.1.2 Go-Back-N ARQ

In contrast to stop-and-wait ARQ, the go-back-N ARQ scheme is continuous, i.e., the packets are transmitted continuously without waiting for any acknowledgements. The packets are sequentially numbered and buffered at the transmitter for possible future use. If the receiver detects errors in a particular packet, a NAK along with the number of the erroneous packet is sent to the transmitter. When the NAK is received by the transmitter, it may be transmitting a packet with a sequence number greater than that of the erroneous packet. The transmitter *goes back* to retransmit the erroneous packet, and every packet after it. Thus, many packets may be sent more than once even if they were received correctly. For a detailed performance analysis of go-back-N ARQ scheme, see [289] and [60]. We summarize the performance based on [285].

Taking the retransmissions into account, the average number of transmissions (original and retransmissions) is

$$\begin{aligned} T_{GBN} &= 1 \cdot P + (N+1)P(1-P) + (2N+1)P(1-P)^2 + \cdots \\ &\quad + (lN+1)P(1-P)^l + \cdots \\ &= 1 + \frac{N(1-P)}{P}. \end{aligned} \tag{9.4}$$

Therefore, the throughput of a go-back-N ARQ scheme is

$$\eta_{GBN} = \frac{1}{T_{GBN}} \left(\frac{k}{n}\right) = \frac{P}{P + (1-P)N} \left(\frac{k}{n}\right). \tag{9.5}$$

The reliability of go-back-N ARQ is given by (9.1).

### 9.1.3 Selective-Repeat ARQ

While the go-back-N ARQ scheme uses the channel continuously and does not spend any time idling, it is still inefficient due to the fact that a packet may be sent more than once even if it was received correctly. The selective-repeat ARQ scheme removes this inefficiency by selectively retransmitting only those packets that were received in error. Correctly received packets are buffered at the receiver until all the erroneous packets are received correctly so that they can be delivered to the user in consecutive order. While this protocol is ideal in principle, it may require infinite buffer storage at the receiver. In practice, however, a finite buffer is used and a time-out mechanism is implemented to limit the waiting time before retransmissions can be initiated for a packet. For a detailed performance analysis of selective-repeat ARQ, see [333], [461], and [11]. We summarize the performance based on [285].

Taking the retransmissions into account, the average number of transmissions (original and retransmissions) is

$$\begin{aligned} T_{SR} &= 1 \cdot P + 2 \cdot P(1-P) + 3 \cdot P(1-P)^2 \\ &\quad + \cdots + l \cdot P(1-P)^{l-1} + \cdots \\ &= \frac{1}{P}. \end{aligned} \tag{9.6}$$

Therefore, the throughput of a selective-repeat ARQ scheme is

$$\eta_{SR} = \frac{1}{T_{SR}} \left(\frac{k}{n}\right) = \left(\frac{k}{n}\right) P. \tag{9.7}$$

The reliability of selective-repeat ARQ scheme is also given by (9.1).

# 9.2 CONVOLUTIONAL TYPE-I HYBRID ARQ SCHEMES

All the commonly used convolutional decoding techniques described in Part III of this book can be modified for use in type-I hybrid-ARQ protocols by incorporating a suitable reliability criterion to determine when to stop decoding a received data packet and request retransmission. For simplicity, the discussion is limited to rate one-half coding.

Assume that an $N$-bit packet is transmitted over a BSC with transition probability $p$ and a rate one-half modified decoder is used. The performance of a type-I hybrid ARQ scheme is characterized by two parameters: reliability and throughput. The reliability is measured in terms of the probability of an accepted packet being in error. The throughput is determined by the average number of retransmissions required to correctly receive a packet. In conventional ARQ, all detectable errors generate a retransmission request. Hence, the reliability is very high and is equal to the probability of undetected error. The reliability of an FEC scheme is $P_b(E)$, the bit error probability after decoding, and the throughput is just the coding rate.

In a hybrid-ARQ scheme, the probability of bit error in an accepted packet is [286]

$$P(E) = P_b(E)(1 + P_r + P_r^2 + ...) = \frac{P_b(E)}{1 - P_r}, \tag{9.8}$$

where $P_r$ is the probability that a single transmission will cause generation of a retransmission request. The throughput of a type-I hybrid-ARQ decoder using ideal selective-repeat protocol is [286]

$$\eta_{SR} = \frac{1}{2}(1 - P_r). \tag{9.9}$$

Note that the throughput for stop-and-wait or go-back-N protocols can be similarly calculated.

## 9.2.1 Modified Viterbi Decoding

The type-I hybrid-ARQ modifications to the Viterbi algorithm suggested in the literature are based on the difference between the path metric of two converging paths [481], an absolute value of the survivor path metric over the entire received packet [463], and the survivor path metric over a fixed size sliding window of the received packet [188, 189]. Related work also appears in [133, 257].

## *Yamamoto and Itoh's Modification*

The modification proposed by Yamamoto and Itoh [481] is based on the difference between the path metric of two converging paths. Recall from Chapter 5 that the Viterbi decoder computes the metric for each path at each level of the trellis for the received sequence. It selects the path with the best metric among those merging at each state of that level, and the other paths are discarded.

At level $m$, all the paths are labeled "reliable". At each node of level $j$ ($j = m+1,\ m+2,\ \ldots$), the paths with the best and the next best metrics are selected. If the path with the best metric is labeled "reliable" and the difference between the metrics of the best and the next best paths is greater than a predetermined threshold (a positive constant), the best path survives with the label "reliable". Otherwise, it survives with label "unreliable". This procedure is continued until either all survivors are labeled "unreliable" at some level, in which case a retransmission request is sent to the transmitter, or the decoding is completed by processing the $m$ termination levels.

## *Wicker's Modification*

The modification proposed by Wicker [463] is based on an absolute value of the survivor path metric over the entire received packet. If the metric $M(\mathbf{r}|\mathbf{v}) = -log\ P(\mathbf{r}|\mathbf{v})$ is selected, the best paths through the trellis have the lowest metric values. The partial path metric $M(\mathbf{r}_{[0,j]}|\mathbf{v}_{[0,j]})$ is the sum of individual bit metrics $M(r_t|v_t)$ up to the $(j+1)$th level in the trellis. In this case, the value of the partial metric $M(\mathbf{r}_{[0,j]}|\mathbf{v}_{[0,j]})$ is directly proportional to the error content of $\mathbf{r}$ at the $(j+1)$th node in the trellis given that $\mathbf{v}$ was transmitted. This suggests the use of an absolute threshold to determine when a path has sufficient error content to be deemed unreliable.

The reliability criterion used is the following [463]: any surviving path whose partial metric achieves a predetermined threshold will be deleted from consideration; if all paths are deleted before decoding is completed, then a retransmission is requested.

## *Harvey and Wicker's Modification*

The modification proposed by Harvey and Wicker [188, 189] is based on the survivor path metric over a fixed size sliding window of the received packet. The Hamming distance between the received sequence and each encoded sequence over a specific interval or window is determined by the growth of the path metric. The number of branches in the window is called the *transition trap length* $J_T$. Consider all paths which merge with the all zero path at a particular node in the trellis. $J_T$ is the minimum number of branches over which all such paths will have a Hamming weight greater than or equal to $d_{free}$. The rate of increase of the path metric over the last $J_T$ branches at each node at level $t$ is defined as the difference between the survivor path metrics at levels $t$ and $t - J_T$.

The following reliability criterion is used [188]: if the rate of increase of the path metric exceeds a fixed retransmission threshold, then the path is considered unreliable and is terminated. If all paths are terminated before completion of the decoding algorithm, then retransmission of the packet is requested.

## *Performance*

The fixed-redundancy hybrid ARQ scheme proposed by Harvey and Wicker [188, 189] is analyzed below. Assume, without loss of generality, that the all-zero encoded sequence was sent. The probability $P_d$ of an incorrect path of weight $d$ being chosen, an event that occurs only if there were $d-\tau+1$ or more errors among the $d$ bits, is given by

$$P_d = \sum_{e=d-\tau+1}^{d} \binom{d}{e} p^e (1-p)^{d-e} \leq 2^d p^{d-\tau+1}, \tag{9.10}$$

where the simplified bound is obtained for small $p$. Using the union bound to overbound the probability of packet error $P_e$, we get

$$P_e < \sum_{d=d_{free}}^{\infty} A_d P_d < \sum_{d=d_{free}}^{\infty} A_d 2^d p^{d-\tau+1} \tag{9.11}$$

$$\approx A_{d_{free}} 2^{d_{free}} p^{d_{free}-\tau+1}. \tag{9.12}$$

By comparing the exponent of $p$ in (9.12) and (5.8), we see that the modified Viterbi algorithm reduces the upper bound in the probability of packet error by a factor of $p^{(d_{free}/2)-\tau+1}$.

Similarly, throughput can be calculated. The probability of successfully receiving the packet up to the $J_T$th transition is

$$P_{s0} = \sum_{j=0}^{\tau-1} \binom{nJ_T}{j} p^j (1-p)^{nJ_T-j}, \tag{9.13}$$

and the probability of successful reception at the $(k + J_T)$th transition is

$$P_{sk} = \frac{1}{P_{s0}} \sum_{i=0}^{min(n,\tau-1)} \binom{n}{i} \sum_{j=0}^{\tau-i-1} \binom{nJ_T - n}{j} \cdot \sum_{l=j}^{min(\tau-1,j+n)} \binom{n}{l-j} p^{l+i}(1-p)^{nJ_T+n-i-l}. \quad (9.14)$$

Finally, the total probability of successful packet transmission is

$$P_s = P_{s0}(P_{sk})^{\frac{N}{n}-J_T}. \quad (9.15)$$

This probability is the conditional probability of decoding a packet successfully at level $J_T + k$, given that all previous levels were successful.

## 9.2.2 Modified Sequential Decoding

In [120, 122], two variants of sequential decoding are proposed that use a time-out condition and the growth rate of the metric as reliability criteria. Another sequential decoding based type-I hybrid-ARQ protocol is discussed in [410].

### *Drukarev and Costello's Modification*

Two algorithms are described by Drukarev and Costello [120, 122]. One is a sequential decoding algorithm with a time-out modification. This algorithm was proposed by Kahn *et al.* [231]. A message is divided in blocks and each block is encoded separately using a convolutional code. At the receiver, each encoded sequence is allowed a fixed maximum time for decoding. That is, the entire encoded sequence must be decoded within this time, or equivalently, must take less than a fixed maximum number of computations for decoding. If each such packet is successfully decoded, it is accepted. Otherwise, a retransmission request is generated for the packet whose decoding exceeds the limit. This algorithm achieves very low error probability but does not attain optimal throughput.

The algorithm proposed by Drukarev and Costello is based on computing the change in the metric value over a fixed number of branches, and requesting a retransmission if this change is below a predetermined threshold. This algorithm is called the *slope control algorithm*. A slowly changing metric is an indication that many paths near the top of the stack have metric values close to the metric value of the top path. All these paths are likely to be extended before a clear winner appears, thereby requiring a large number of computations per branch. The strength of the slope control algorithm is in predicting the approach of the time-out and requesting a retransmission earlier than the time-out algorithm does.

## *Shacham's Modification*

Shacham [409, 410] has analyzed the performance of a hybrid ARQ scheme with sequential decoding and the time-out reliability criterion. Optimum values for time-out duration are obtained for one-hop and two-hop radio links [409]. Queueing is allowed at the receiver to model the situation where the decoder speed is not fast enough to keep up with the incoming packets and a queue is built up. Average queue lengths and rate of packet arrivals that can be supported by such a system are analyzed, and optimum time-out duration is determined by taking the decoder's queue into consideration.

## *Performance*

Drukarev and Costello [120, 122] have analyzed the performance of the slope-control algorithm. Their results are summarized here. For every node $j$ on the correct path, the $j$th retransmission subset is defined to be the incorrect subset for that node plus the node itself. The probability that a retransmission request is generated in the $j$th retransmission subset is upper bounded by

$$P_r \leq Ge^{\beta\Gamma}, \tag{9.16}$$

where $G$ is a constant that depends on channel conditions, the transmission rate, and the window length, $\beta < \frac{\rho}{\rho+1}$ with $\rho$ a constant that depends on the channel conditions, $\Gamma$ is a predetermined threshold used in requesting retransmissions, and $R \leq (1-\epsilon)\frac{E_0(\rho)}{\rho}$, $0 < \rho < 1$, with $\epsilon$ a positive constant smaller than 1.

The undetected bit error probability for sequential decoding is upper bounded by $P_e \leq D2^{-K\gamma}$, where $D$ and $\gamma$ are constants that depend on the channel conditions and transmission rate, and $K$ is the input constraint length of the encoder. The bit error probability of the slope control algorithm is upper bounded by

$$P_e \leq D2^{-K\gamma}e^{-\alpha(\Gamma-\Gamma_{min})}(1-sGe^{\beta\Gamma}), \tag{9.17}$$

where $\alpha = \frac{(1-\epsilon)\rho^2}{1+\rho}$, $0 < \rho < 1$, $\Gamma_{min}$ is the minimum change in the metric over a window of $W$ branches.

### 9.2.3 Modified Majority-Logic Decoding

The majority-logic decoding algorithm can also be modified for use in a hybrid-ARQ protocol [466] by favoring a strong majority and by labeling a weak majority as unreliable. Recall from Chapter 7 that an FEC majority-logic decoder estimates an information error bit $e$ as 1 if more than half of the $J$ check sums orthogonal on $e$ have a value of one; otherwise, $e$ is estimated as zero. This majority-logic decoding rule can be modified in a simple way to furnish reliability information that can be exploited in a hybrid ARQ scheme.

Let $\eta$ be the number of check sums that have a value of one. Let $\tau$ be an integer less than $J$. The modified majority-logic rule for estimating the value of $e$ is as follows:

$$\begin{cases} \eta \geq \lfloor \frac{J}{2} \rfloor + \lfloor \frac{\tau}{2} \rfloor + 1 : & e = 1 \\ \lfloor \frac{J}{2} \rfloor - \lfloor \frac{\tau}{2} \rfloor < \eta < \lfloor \frac{J}{2} \rfloor + \lfloor \frac{\tau}{2} \rfloor + 1 : & \text{RequestRetransmission} \\ \eta \leq \lfloor \frac{J}{2} \rfloor - \lfloor \frac{\tau}{2} \rfloor : & e = 0 \end{cases} .$$

The most frequent error patterns result in almost all of the $J$ check sums to be either one or zero and the decoding proceeds normally. The less frequent error patterns result in a less pronounced majority among the $J$ check sums and are most likely to cause decoder errors. Occurrence of these error patterns triggers retransmission requests.

## *Performance*

The probability $P_e$ of a particular information bit being decoded incorrectly by the modified majority-logic decoder on any given transmission is overbounded as follows:

$$P_e \leq \sum_{e=t_{ML}+\lfloor \frac{\tau}{2} \rfloor+1}^{n_E} \binom{n_E}{e} p^e(1-p)^{n_E-e} \approx C_{mm} \; p^{t_{ML}+\lfloor \frac{\tau}{2} \rfloor+1}, \qquad (9.18)$$

where the approximation is for a small value of $p$ and $C_{mm}$ is a constant dependent on the structure of the encoder. A more detailed computation of this bound is given in [466]. As in the case of modified Viterbi decoding, the modified majority-logic algorithm reduces the upper bound on the probability of bit error on any given transmission, in this case by a factor of $p^{\lfloor \frac{\tau}{2} \rfloor}$.

An upper bound on the probability of retransmission $P_r$ is obtained by using the union bound and is given by

$$P_r \leq \frac{N}{2} \sum_{e=t_{ML}-\lceil \frac{\tau}{2} \rceil+1}^{n_E} \binom{n_E}{e} p^e(1-p)^{n_E-e}, \tag{9.19}$$

where $N/2$ error estimates are made in an $N$ bit packet.

### 9.2.4 Modified Table Look-Up Decoding

The table look-up decoding described in Chapter 8 can be modified for use in a type-I hybrid-ARQ protocol [112]. We have seen that the correction table in a table look-up decoder can be set up to guarantee correction of up to and including all $t_j$-bit errors in a block of $2(j+1)$ bits. This is because there are no conflicts in the correction table for error patterns with $t_j$ or fewer errors. In addition, a table look-up FEC decoder can also correct many $i$-bit errors where $t_j + 1 \leq i \leq t_j + \tau$. It is possible to improve the probability of decoder error if the FEC scheme is modified to request retransmission for errors that cannot be corrected but are detected.

The correction table for the modified table look-up decoding scheme is constructed to flag detectable but uncorrectable errors. These are the error patterns that generate a table conflict. The table generation proceeds as in the table look-up FEC scheme of Chapter 9, starting with single-bit errors and moving up to higher order errors. Addresses that correspond to conflicts are marked for retransmission. After all the patterns with up to $t_j + \tau$ bits in error have been processed, some table entries may still be empty. These addresses may be generated by patterns with more than $t_j + \tau$ bits in error. These empty addresses should also be marked for retransmission.

The modified table look-up decoding is the same as the simple table look-up decoding if no s-addresses with conflicts are generated in decoding a packet. The only difference is that the correction table has an extra bit in each entry, identifying the s-addresses corresponding to conflicts. The decoding of a packet is stopped if an s-address with conflicts is generated, and a retransmission request is issued for the packet.

### *Performance*

The modified table look-up decoding allows correction of all $i$-bit errors, $0 \leq i \leq t_j + \tau$, either in the first transmission on in subsequent retransmissions. Hence, the probability of decoder error for the table look-up type-I hybrid-ARQ protocol is

$$P_b(E) \leq \frac{1}{2} \sum_{i=t_j+\tau+1}^{2(j+1)} \binom{2(j+1)}{i} p^i (1-p)^{2(j+1)-i} \tag{9.20}$$

$$\approx \frac{1}{2} \binom{2(j+1)}{t_j+\tau+1} p^{t_j+\tau+1} = C_{mt} p^{t_j+\tau+1}, \tag{9.21}$$

where the approximation is for small values of $p$ and $C_{mt}$ is a constant dependent on the structure of the encoder and the correction table. Note that the exponent of $p$ in (9.21) is larger by $\tau$ compared to (8.9), thereby reducing the probability of decoder error.

In modified table look-up decoding, retransmission requests are generated by errors that the table flags but does not correct. The probability of a retransmission request being generated in the first $2(j+1)$-bit block of a received packet is

$$P_{r1} \leq \sum_{i=t_j+1}^{t_j+\tau} \beta_i \binom{2(j+1)}{i} p^i (1-p)^{2(j+1)-i}, \tag{9.22}$$

where a retransmission request is assumed to be generated by $i$-bit error patterns with $t_j + 1 \leq i \leq t_j + \tau$ for which table indicates a conflict, and $\beta_i$ is the fraction of such error patterns. In an $N$ bit packet, $N/2$ table look-ups are made. The probability that no retransmission requests will be generated for a packet is

$$(1 - P_{r1})^{\frac{N}{2}}, \tag{9.23}$$

and the probability that at least one retransmission request will be generated is

$$P_r = 1 - (1 - P_{r1})^{\frac{N}{2}}. \tag{9.24}$$

## 9.3 METHODS OF PERFORMANCE IMPROVEMENT

In this section, some common approaches for improving the performance of fixed-redundancy error control schemes are discussed.

## *Multiple Copy*

In a variation of selective-repeat ARQ suggested by Weldon [461], a packet is multiply retransmitted in increasing numbers, as more retransmissions become necessary, to improve the chances of at least one copy being received correctly.

## *ARQ with Memory and Code Combining*

In type-I hybrid ARQ, each retransmission is processed independently. If uncorrectable errors are detected in a packet, it is completely discarded and a retransmission is requested. However, it is quite likely that many bits in the discarded packet are correct and contain useful information. Thus, storing a packet with errors at the receiver and combining it in some way with the retransmitted copies of the packet may be useful. This observation has led to the development of ARQ schemes with memory. One such scheme was proposed by Sindhu [422].

Benelli [29] proposed an ARQ scheme with memory and soft error detectors. The receiver stores all copies of the received codeword and combines them using soft decision reliability information. This process automatically leads to the correction of errors as the reliability estimate improves with time. Improvements to this ARQ with memory scheme by integrating modulation with channel coding are described in [30, 31].

Code combining, first suggested by Chase [74, 75], is a technique for combining codewords of progressively lower rate by combining unreliable retransmissions in an optimum manner. Use of code combining results in an adaptive type-I hybrid ARQ scheme that has a more graceful throughput degradation. Kallel and Haccoun [240] have proposed an adaptive type-I hybrid ARQ scheme with code combining. The underlying type-I hybrid ARQ uses convolutional coding and sequential decoding with a time-out condition as a criterion for stopping the decoding process and requesting retransmission. Kallel and Haccoun [235] have suggested an improvement based on partial retransmissions which makes a better use of the channel bandwidth.

Wicker has also proposed an adaptive type-I hybrid ARQ schemes using diversity combining and majority-logic decoding [464, 467] and another scheme using Reed-Solomon codes [465, 467]. His scheme guarantees that an odd number of copies are available at the receiver at any time and a majority voting scheme can be used to determine a reliable single copy which is then used in a correction scheme. This allows the resulting scheme to adapt to the channel noise rate. Finally, variable-redundancy error control schemes can be characterized as being based on code combining and are discussed separately in Chapter 10.

### *Information Feedback*

The NAK response sent to the transmitter to request retransmission can be modified to carry some information about the nature of the errors detected at the receiver in the previous transmission. This technique is called information feedback. Arazi [18] has proposed an improvement to stop-and-wait ARQ by using information feedback.

## 9.4 EXERCISES

1. Assume that a (511, 484) block code is used on a BSC with transition probability $p = 0.005$ for error detection in a stop-and-wait ARQ scheme. Compute the error probability $P(E)$.

2. The error-detecting code and the BSC of Exercise 1 are used in a go-back-N ARQ. Compute the throughput of this scheme, given that $N = 10$.

3. Repeat Exercise 2 for a selective-repeat ARQ scheme.

4. The approximate probability of packet error in modified Viterbi decoding is given by (9.12). Compute this probability for the (2, 1, 2) encoder with generator sequences $\mathbf{g}^{(1)} = 101$ and $\mathbf{g}^{(2)} = 111$ that has $d_{free} = 5$ and $A_{d_{free}} = 1$. Assume that the BSC of Exercise 1 is used.

5. The approximate probability of packet error in modified majority-logic decoding is given by (9.18). Compute this probability for a (2, 1, 35) encoder with majority-logic error-correcting capability $t_{ML} = 4$. Assume that $n_E \approx n_A = n(m+1)$ for this encoder and the BSC of Exercise 1 is used.

6. For the encoder and BSC given in Exercise 5, find the probability of retransmission using (9.19).

# 10

# VARIABLE-REDUNDANCY ERROR CONTROL SCHEMES

Conventional ARQ schemes employ a high-rate code for error detection only. This maximizes the throughput during good channel conditions. However, their throughput degrades rapidly as the channel conditions deteriorate. This degradation can be postponed by embedding FEC in the protocol (e.g. hybrid-ARQ) to correct common errors, and by requesting retransmissions only when uncorrectable errors are detected. The net effect of using a more powerful code for both error detection and correction is an increase in the amount of redundancy in the transmitted message, which reduces the maximum throughput achievable. Conventional ARQ, FEC, as well as hybrid-ARQ are essentially fixed-redundancy schemes. Nonstationary channels such as those encountered in digital mobile communications require adaptive error control. Fixed-redundancy schemes are not suitable for use on these channels since the the coding redundancy is either "too much" under good channel conditions, or "too little" under bad channel conditions, wasting bandwidth in the former case and increasing retransmission frequency in the latter.

*Variable-redundancy* hybrid-ARQ (VR-HARQ) schemes are designed to vary the amount of coding redundancy in accordance with the channel conditions. The basic idea is to send parity check bits for error correction to the receiver only when they are needed. Thus, VR-HARQ schemes behave like conventional ARQ under good channel conditions and, as the channel conditions deteriorate and FEC is initiated, become progressively better hybrid-ARQ schemes.

## 10.1 CLASSIFICATION OF VR-HARQ SCHEMES

A significant amount of research effort has been devoted over the past two decades to the design and development of VR-HARQ schemes. A simple mechanism can be identified that is common to all VR-HARQ schemes. The essential steps in this *basic* VR-HARQ scheme are as follows. It is just like conventional ARQ in the first transmission. The second transmission, if required, consists of parity symbols generated using a more powerful error-correcting code.

The retransmission strategy used when more than two transmissions are required for a message determines further classification of the schemes. We use two classes of VR-HARQ scheme. One class of schemes, which we will call *class A* VR-HARQ, is based on type-II hybrid ARQ suggested by Lin and Yu [291] and later modified by Wang and Lin [288, 460]. Their scheme involves alternate transmissions of information and parity blocks.

In contrast to class A VR-HARQ, a scheme proposed by Hagenauer [176] uses successive parity transmissions such that a progressively lower rate code is built up until it is powerful enough to decode the message. Such a scheme was originally suggested by Mandelbaum [304] using block codes and by Ancheta [12] using convolutional codes. Hagenauer's scheme is the basis of *class B* VR-HARQ schemes.

### 10.1.1 Basic VR-HARQ Schemes

The basic VR-HARQ scheme is limited to two transmissions per message. Two levels of redundancies are used in this scheme; one of them is based on a high-rate error-detecting code similar to conventional ARQ, and the second is based on the the error-correcting code. The difference between this and a type-I hybrid-ARQ scheme lies in the use of the first level of redundancy which makes this scheme perform better under good channel conditions by maintaining a very high throughput. Examples of basic VR-HARQ schemes can be found in [101, 327].

Let $C_1$ be a high-rate $(n, k)$ block code that is used for error detection only. Let $C_2$ be a (2, 1, $m$) convolutional code. Further, let $\mathbf{w}$ be a $k$ bit message to be transmitted. This message is encoded using $C_1$ into $n$-bit sequence $\mathbf{u}$. Next, an encoder for the code $C_2$ is used to encode $\mathbf{u}$ into a $2(n+m)$-bit codeword $\mathbf{v}$. Note that $m$ zeros are used to terminate the convolutional code. The $2(n+m)$-

bit codeword $\mathbf{v}$ is separated into the two $(n+m)$-bit encoded sequences $\mathbf{v}^{(1)}$ and $\mathbf{v}^{(2)}$. Since $C_2$ is a convolutional code, it is possible to recover $\mathbf{u}$ either from $\mathbf{v}^{(1)}$ or from $\mathbf{v}^{(2)}$ by dividing by the appropriate generator polynomial.

Initially, $\mathbf{v}^{(2)}$ is stored in a transmitter buffer, and $\mathbf{v}^{(1)}$ is sent to the receiver as the first transmission. Let the received sequence corresponding to $\mathbf{v}^{(1)}$ be $\tilde{\mathbf{v}}^{(1)}$. The receiver divides $\tilde{\mathbf{v}}^{(1)}$ by $\mathbf{g}^{(1)}$ to recover $\tilde{\mathbf{u}}$. A syndrome sequence is computed using the code $C_1$ for error detection. If the syndrome is zero, the received block is assumed to be error-free. It is decoded using $C_1$ to extract the original message $\tilde{\mathbf{w}}$ which is delivered to the user. An ACK is sent to the transmitter. If, however, the syndrome is nonzero, error are said to have occurred in the first transmission, and a NAK is sent to the transmitter. The erroneously received block $\tilde{\mathbf{v}}^{(1)}$ is stored in the receiver buffer for possible future use.

Upon receiving the NAK, the transmitter responds by retrieving $\mathbf{v}^{(2)}$ from the buffer and sending it to the receiver. Let the received block corresponding to $\mathbf{v}^{(2)}$ be $\tilde{\mathbf{v}}^{(2)}$. The receiver divides $\tilde{\mathbf{v}}^{(2)}$ by $\mathbf{g}^{(2)}$ to recover $\tilde{\mathbf{u}}$, and again computes a syndrome sequence using the code $C_1$ for error detection. If the syndrome is zero, the received block is assumed to be error-free. It is decoded using $C_1$ to extract the original message $\tilde{\mathbf{w}}$ which is delivered to the user. An ACK is sent to the transmitter, and the previously stored $\tilde{\mathbf{v}}^{(1)}$ is discarded. If, however, the syndrome is nonzero, errors are detected in the second transmission and the receiver attempts error correction. This is done by combining $\tilde{\mathbf{v}}^{(1)}$ and $\tilde{\mathbf{v}}^{(2)}$ into a single $2(n+m)$-bit composite sequence $\tilde{\mathbf{v}}$. A decoder for the code $C_2$ is used to correct the errors and decode the original message. Let the decoded message be $\tilde{\mathbf{u}}$. In a basic scheme, $\tilde{\mathbf{u}}$ is decoded using the code $C_1$ to get $\tilde{\mathbf{w}}$, which is delivered to the user.

### 10.1.2 Class A VR-HARQ Schemes

In class A VR-HARQ, the first two transmissions corresponding to a message are the same as those in basic VR-HARQ, with the difference that both the information and parity blocks are stored in the transmitter buffer for possible future use. All retransmissions consist of alternate information and parity blocks. The receiver uses the most recent copies of the information and parity blocks together to correct errors using the error-correcting code. Examples of class A VR-HARQ are proposed in [291, 288, 460, 297, 298, 87, 88, 468, 244, 116].

Having described the basic VR-HARQ scheme in the previous section, we now outline the additional steps below. Let the decoded message after two transmissions be $\tilde{\mathbf{u}}$. This message is checked for correctness by computing a syndrome using $C_1$. If the syndrome is zero, the $n$-bit sequence $\tilde{\mathbf{u}}$ is assume to be error-free. The original message $\tilde{\mathbf{w}}$ is extracted by decoding using $C_1$ and delivered to the user, and an ACK is sent to the transmitter. If the syndrome is not zero, the receiver sends a NAK to the transmitter, and stores $\tilde{\mathbf{v}}^{(2)}$ in place of $\tilde{\mathbf{v}}^{(1)}$ in its buffer.

The transmitter responds to the second NAK by retransmitting $\mathbf{v}^{(1)}$, which is received as $\tilde{\mathbf{v}}^{(1)}$ by the receiver. The information sequence $\tilde{\mathbf{u}}$ is recovered from it and a syndrome is computed on $\tilde{\mathbf{u}}$ using $C_1$. If the syndrome is zero, $\tilde{\mathbf{u}}$ is assumed error-free and is decoded using $C_1$ and the decoded message is deliver to the user. An ACK is sent to the transmitter. If the syndrome is not zero, $\tilde{\mathbf{v}}^{(1)}$ is combined with the previously store $\tilde{\mathbf{v}}^{(2)}$ and error correction using $C_2$ is attempted on the composite sequence $\tilde{\mathbf{v}}$. Let the decoded message be $\tilde{\mathbf{u}}$. This message is checked for correctness by computing a syndrome using $C_1$. If the syndrome is zero, the $n$-bit sequence $\tilde{\mathbf{u}}$ is assumed to be error-free. The original message $\tilde{\mathbf{w}}$ is decoded by using $C_1$ and delivered to the user, and an ACK is sent to the transmitter. If the syndrome is not zero, the receiver sends a NAK to the transmitter, and stores $\tilde{\mathbf{v}}^{(1)}$ in place of $\tilde{\mathbf{v}}^{(2)}$ in its buffer. The entire process is then repeated until the message is received without errors.

### 10.1.3 Class B VR-HARQ Schemes

Class B VR-HARQ schemes differ from class A VR-HARQ schemes in that each retransmission consists of additional parity check bits. The first transmission consists of the original message appropriately encoded using a high-rate code for error detection. The first retransmission (second transmission) consists of parity-check bits obtained by using a more powerful error-correcting code. All subsequent retransmissions carry parity-check bits obtained by using a progressively lower rate code. The basic idea is to continue lowering the code rate by transmitting additional parity bits until the code is powerful enough to correctly decode the message. Examples of class B VR-HARQ are proposed in [304, 12, 176, 241, 235, 237].

The basic VR-HARQ scheme can be extended further to a class B scheme. The first two transmissions in a Class B scheme are processed as in a basic scheme. The third transmission (second retransmission) is made up of additional parity bits, obtained by using a lower rate encoder. The receiver attempts decoding using all of the available redundancy, i.e., using the lower rate codeword. The decoded output is checked for errors using the error-detecting code $C_1$. If no errors are detected, the decoded message is accepted after stripping the parity-check bits, and an ACK is sent to the transmitter. If, however, errors are still detected in the decoded block, a NAK is sent. The transmitter sends yet another block of parity bits, obtained by using an even lower rate encoder, and the process continues.

In the schemes suggested by Hagenauer [176, 178], and Kallel and Haccoun [241], rate-compatible encoders are used, along with a Viterbi decoder for the original rate-1/2 encoder. Thus, a class B scheme accommodates incremental redundancy in successive retransmissions.

### 10.1.4 Type of Coding Used in VR-HARQ Schemes

Variable-redundancy hybrid ARQ schemes suggested in the literature have been based on block as well as convolutional codes. All the schemes use a high-rate block code for initial error detection. The second, more powerful, error-correcting code can be block or convolutional. The use of block codes in VR-HARQ has been proposed in [101, 304, 327, 291, 460, 328, 253, 125, 338, 252, 283, 468]. The use of convolutional codes in VR-HARQ has been proposed in [12, 327, 288, 297, 460, 76, 176, 298, 241, 234, 341, 235, 237, 116].

## 10.2 VR-HARQ SCHEMES USING CONVOLUTIONAL CODES

The VR-HARQ schemes proposed in the literature have used many different classes of convolutional encoders. Encoders with special structural properties have allowed improvements in performance as well as reduction in instrumentation complexity. The use of some of these classes of encoders is illustrated in the following sections.

### 10.2.1 RCC Encoders

Rate-compatible convolutional (RCC) encoders were defined in Chapter 4. The RCC encoders can be used to provide incremental redundancy, i.e., for providing redundant symbols corresponding to a message gradually over multiple transmissions. All the redundant symbols may be computed before any transmission begins and divided into subsequences which are stored in the transmitter buffer. Each retransmission request results in a new subsequence being sent to the receiver. The redundant symbols may alternately be computed only when they are requested, thereby saving buffer space at the transmitter. The reader may refer to [178, 180, 241] for details.

### 10.2.2 Locally Invertible Encoders

The $(2, 1, m)$ locally invertible encoders were defined in Section 4.6.1 and were shown to have the property (see Property 4.3) that a composite encoded sequence $\mathbf{v}$ can be split into two sequences $\mathbf{v}_I$ and $\mathbf{v}_{II}$ such that any of these two can be inverted to obtain the information sequence $\mathbf{u}$. Thus, each subsequence performs two tasks: it carries all the information necessary to recover the original message in case of an error-free transmission; it also acts as additional parity information which can be used to perform error correction in case of an erroneous transmission. For details, see [116]. The use of invertible rate one-half block codes in a type-II hybrid-ARQ scheme was proposed in [285, 460].

### 10.2.3 Performance

The performance of VR-HARQ scheme is characterized by its throughput and reliability. The throughput is evaluated by computing the probability of retransmission, and the reliability is the probability that an accepted decoded packet contains one or more symbol errors. We assume that selective-repeat ARQ scheme is being used. The analysis can be easily extended to other types of ARQ schemes. We also assume that the transmission channel can be modeled by a BSC with bit error probability $p$, and that the feedback channel is error-free. While these assumptions are not realistic, they simplify analysis. The reader may refer to [460, 88, 178, 241, 484] for further details.

## *Probability of Retransmission and Throughput*

The probability that a block of $n+m$ bits will be received correctly is $P_c = (1-p)^{n+m}$. Let $P_b$ be the bit error probability after decoding based on the convolutional code $C_2$ and a Viterbi decoder. Let $T(X, Y)$ be the generating function of the code $C_2$. Then, $P_b$ is upper bounded by

$$P_b \leq \frac{\partial}{\partial Y} T(X, Y) \left|_{X=\sqrt{2p(1-p)},\, Y=1} \right. . \tag{10.1}$$

Let $q = (1-P_b)^{n+m}$. Let $P_t = P_c + (1-P_c)q$. Let $N$ be the number of $(n+m)$-bit subsequences that can be transmitted in a round-trip delay. Let $\eta_\infty$ be the throughput efficiency of the VR-HARQ scheme, assuming that the receiver has an infinite buffer. Then, we have

$$\eta_\infty \geq \frac{1+(1-P_c)P_t/P_c}{2-P_c} \cdot \frac{n}{n+m} \cdot \eta_0, \tag{10.2}$$

where $\eta_0 = (k/n)P_c$ is the throughput efficiency of the pure selective-repeat ARQ with infinite receiver buffer.

## *Probability of Error*

Let $P_e$ be the probability of an undetected error for the $(n,\ k)$ block code $C_1$. Let $P(E)$ be the probability that the decoder of the VR-HARQ system commits a decoding error. It can be shown that

$$P(E) \leq \frac{2P_e}{P_e + P_c}. \tag{10.3}$$

If we choose $C_1$ properly, $P_e$ is upper bounded by

$$P_e \leq [1-(1-p)^k]2^{-(n-k)}. \tag{10.4}$$

We can make $P_e$ very small by selecting a moderate number of parity bits, $(n-k)$, in comparison to $P_c$, and thus make $P(E)$ very small. This makes the system very reliable, with a probability of error of the same order as that of pure selective-repeat ARQ.

# 10.3 EXERCISES

1. Consider a VR-HARQ scheme with the following parameters: a $(511, 493)$ block code is used on a BSC with transition probability $p = 0.001$ for

error detection in a selective-repeat ARQ scheme, and a (2, 1, 2) convolutional encoder with $d_{free} = 5$ and $A_{d_{free}} = 1$ is used for error correction. Compute the error probability $P(E)$ of this scheme.

2. For the VR-HARQ scheme of Exercise 2, determine the throughput, given that the underlying selective-repeat ARQ uses an infinite buffer.

# 11

# ERROR RECOVERY IN HIGH-SPEED NETWORKS

The performance of high-speed networks of today and the future will be primarily limited by packet loss due to buffer overflows during congested traffic conditions. A message or a *protocol data unit* (PDU) is made up of many smaller packets, and any one of the packets being lost results in the entire PDU being retransmitted. The performance of end-to-end protocols degrades rapidly if error recovery is based on packet loss detection and retransmission of PDUs, i.e., on some form of ARQ. This phenomenon is more pronounced in high-speed networks since the ratio of packet transmission time to propagation delay is very small. Hence, efficient protocols must be designed to provide reliable communication using error control techniques to recover lost packets. Using FEC to reduce the retransmission frequency is an attractive solution for error recovery in high-speed networks.

The basic idea in using FEC for lost packet recovery is to introduce sufficient redundancy in the message prior to transmission so that a lost packet can be reconstructed at the receiver. Adding redundancy results in additional packets being fed to the network. Hence, using FEC is advantageous only if the reduction in the packet loss rate due to coding more than compensates for the increase in the probability of packet loss due to a larger initial number of packets transmitted. Use of FEC in error recovery in high-speed networks is actively being researched [20, 49, 347, 412]. Most of the schemes proposed use some form of block coding. This chapter describes a convolutional coding technique that allows recovery of consecutive lost packets in a high-speed network [55, 113, 114]. A section summarizing the approaches based on block coding is also included.

## 11.1 ATM NETWORKS

Recent advances in high-speed networking technology and the associated demand for flexibility have resulted in the emergence of *asynchronous transfer mode* (ATM) as the ultimate solution for *broadband integrated services digital network* (B-ISDN). The B-ISDN protocol model for ATM has a layered architecture with the following three layers: the physical layer (PHY) which is responsible for transporting information, the ATM layer which is responsible for switching/routing and multiplexing, and the ATM adaptation layer (AAL) which is responsible for adapting the user information to the ATM stream [366]. The AAL has two sublayers: segmentation and reassembly (SAR) and convergence sublayer (CS).

At the transmitter, AAL accepts the user message in the form of a user PDU and produces a sequence of segments which form the payload of ATM cells[1] [57]. Overhead information including sequence numbers is appended to these segments by the AAL. The ATM layer appends a 5 byte header to the 48 byte payload to form a 53 byte cell which is then transmitted over the network. At the receiver, the AAL reassembles the PDU using the payloads of the received ATM cells. A lost cell shows up as a gap in the sequence number space. The decoding step within the AAL attempts to recover the lost cell before delivering the received PDU to upper layers.

For B-ISDN using ATM, the choice of coding scheme depends upon the location of the encoding/decoding operation within the ATM adaptation layer (AAL). Block codes are a good alternative when coding is applied at the SAR sublayer of the AAL or at the ATM layer. Additional parity cells are added to the message to be transmitted. Thus the ATM cell size dictates the block length of the block coding scheme. Convolutional codes are suitable for encoding at the CS of the AAL as shown in Figure 11.1. The entire PDU is encoded within CS prior to segmentation.

## 11.2 SUMMARY OF RELATED WORK

Four basic types of schemes are summarized in this section.

[1] Packets in an ATM network are called cells.

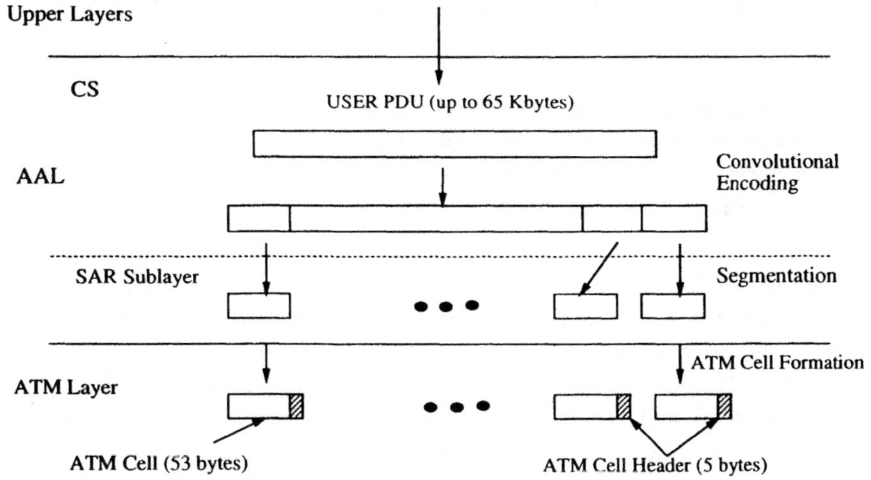


**Figure 11.1** Convolutional encoding and segmentation into ATM cells.

- Scheme A: A simple scheme for recovering a single lost packet in a block of $K$ packets [412, 20]. A parity packet, generated by computing a modulo-2 sum of the $K$ data packets is added to the block prior to transmission.

- Scheme B: Extension of scheme A for recovering two lost packets in a block of $K$ packets by adding two linearly independent parity packets to it [412, 20].

- Scheme C: A generalization of schemes A and B for recovering multiple lost packets. The data packets are arranged in an $M-1 \times N-1$ matrix, to which a row of column-wise parity packets and a column of control packets are added, making the encoded block size equal to $MN$ packets [347]. This extension was also suggested in [412].

- Scheme D: A scheme based on using a more powerful coding technique such as Reed-Solomon coding [49, 248, 323, 492]. This allows recovery of multiple lost packets, and efficient VLSI hardware designs for Reed-Solomon decoders are available.

These lost packet recovery schemes can be compared based on the following criteria:

- Coding efficiency, i.e., the amount of redundancy introduced by coding.

- Hardware implementation complexity of the encoder and decoder.
- Decoding delay at the receiver.
- Distribution of lost packets within the block, i.e., whether the lost packets are located randomly or are consecutive within a block. This determines the necessary "guard" space around the lost packets that corresponds to the packets that are received successfully.

The coding efficiency is determined by network parameters. If a packet arrival rate at a node is $\lambda$, and a single parity packet is added to a block of $K$ data packets, the effective arrival rate after coding becomes $\lambda\frac{K+1}{K}$. Similarly, in scheme C, $(M + N - 1)$ parity/control packets are added to $(M - 1)(N - 1)$ data packets, and the effective packet arrival rate becomes $\lambda\frac{NM}{(M-1)(N-1)}$. If the effective packet arrival rate increases to the point where the network load becomes heavy and queueing delays within the network become excessive, coding ends up increasing the packet loss rate. Table 11.1 summarizes the range of coding rates reported in the literature for these schemes.

| | |
|---|---|
| Scheme A | 6/7, 8/9, 10/11, 19/20, 70/71 [412]<br>255/256 [20] |
| Scheme B | 254/256 [20] |
| Scheme C | 12/20 (=3/5), 285/320 (= 57/64) [347]<br>246/256 [20] |
| Scheme D | 45/50,40/50,35/50 (= 9/10,4/5,7/10) [49, 50]<br>(434,429)(255,233) RS code [492] |

**Table 11.1** Comparison of coding efficiency (rates) in various lost packet recovery schemes.

With respect to the hardware complexity, the simplest scheme is the one in which a parity packet is generated by adding (modulo-2) all the data packets, allowing recovery of a single lost packet. The encoder and the decoder both require a modulo-2 adder of size equal to the packet length. Some additional control hardware may be required as well. Recovery of more than one lost packets requires additional (linearly independent) parity packets, and the hardware complexity of the scheme goes up. Scheme C, which uses a two-dimensional matrix of packets, requires a parity generator for each column. Additional hardware is needed for extracting the control information from each packet and comparing it with the control packets within the matrix. The entire set of packets in the matrix need to be buffered before recovery can begin. The hardware of a Reed-Solomon decoder is even more complicated [323].

The decoding delay is equal to the time it takes to receive all the packets in the block, plus processing time required by the decoder. An entire block of packets, including data, parity, and control packets, has to be buffered before recovery can begin. Thus, significant amount of buffer space is needed. For example, in scheme C, with a matrix size of $16 \times 20$, there are 320 ATM cells, each of size 384 bits, requiring 122880 bits of fast buffer storage.

Another important aspect of these schemes is the distribution of lost packets within the block. The single packet recovery of scheme A can recover any packet within the block. If two types of parities are used, any two packets within the block can be recovered. Consider the matrix-based scheme C, with an $M \times N$ matrix of packets. It can recover a single lost packet in any column, or a sequence of $l$ consecutive lost packets as long as $l < N - 1$. If a column has two lost packets, they cannot be recovered unless a more powerful coding technique is used. Thus, the distribution of lost packets in the matrix has to be well-behaved in order to guarantee full recovery.

## 11.3 ERROR RECOVERY USING CONVOLUTIONAL CODES

An interesting application of convolutional codes generated by locally invertible encoders is described in this section. This application has been proposed recently [114, 113]. Much research remains to be done. It is hoped that the ideas presented in this section would encourage the readers to further investigate applications of convolutional codes for error recovery in high-speed networks.

### 11.3.1 Locally Invertible Encoders

In Chapter 4, locally invertible encoders were described and were shown to have the property that overlapping encoded subsequences correspond to overlapping information subsequences. Furthermore, if the one-to-one mapping length of these subsequences is $w$ and the code rate is $k/n$, then encoded subsequences overlapping in $(w-n)$ bits correspond to information subsequences overlapping in $(w-k)$ bits.

**Example 11.1** Consider the (3, 2, 1) encoder with $K = 4$ and generator sequences $G^{(1)} = 0111$, $G^{(2)} = 1011$, and $G^{(3)} = 0001$. It is locally invertible with a one-to-one mapping length $w = 6$. The corresponding reduced encoding and inverse decoding matrices are, respectively,

$$[\mathbf{G}]_{6\times 6} = \begin{bmatrix} 0 & 1 & 0 & 0 & 0 & 0 \\ 1 & 0 & 0 & 0 & 0 & 0 \\ 1 & 1 & 0 & 0 & 1 & 0 \\ 1 & 1 & 1 & 1 & 0 & 0 \\ 0 & 0 & 0 & 1 & 1 & 0 \\ 0 & 0 & 0 & 1 & 1 & 1 \end{bmatrix}, \text{ and } [\mathbf{G}]_{6\times 6}^{-1} = \begin{bmatrix} 0 & 1 & 0 & 0 & 0 & 0 \\ 1 & 0 & 0 & 0 & 0 & 0 \\ 0 & 0 & 1 & 1 & 1 & 0 \\ 1 & 1 & 1 & 0 & 1 & 0 \\ 1 & 1 & 1 & 0 & 0 & 0 \\ 0 & 0 & 0 & 0 & 1 & 1 \end{bmatrix}.$$

An information subsequence of length $w$ bits is encoded to obtain the corresponding encoded subsequence of length $w$ bits by using (4.10), and the the corresponding inversion is given by (4.11). Let the message to be transmitted be $\mathbf{u}_{[0,7]} = 00011010\ 11011100$, which is encoded using (4.10) to obtain the composite encoded sequence $\mathbf{v}_{[0,6]} = 111010100\ 011001101110$. Property 4.4 of overlapping inverted subsequences can be easily verified. For instance, the encoded subsequence $\mathbf{v}_{[1,2]} = 010100$ is inverted to get $\mathbf{u}_{[1,3]} = 011010$. Shifting in three new encoded bits gives the encoded subsequence $\mathbf{v}_{[2,3]} = 100011$, which is inverted to get $\mathbf{u}_{[2,4]} = 101011$. Note that there is an overlap of $w - 2 = 4$ bits between successive inverted information subsequences. That is, the last four bits of $\mathbf{u}_{[1,3]}$ and the first four bits of $\mathbf{u}_{[2,4]}$ are the same.

## 11.3.2 Recovery of a Single Lost Packet

In this chapter, we only consider rate-$k/n$ locally invertible encoders, with $k = n - 1$ and with constraint length $\nu = lk$ for some integer $l$. From (4.8), the one-to-one mapping length is $w = n\nu$. The original message is encoded prior to segmentation. Each segment is encapsulated with a header and/or trailer that includes the sequence number. Thus, an encoded message traverses the network in the form of multiple packets. Each packet has a sequence number associated with it, which is used by the receiver to ensure that all packets corresponding to a particular encoded message have been received. Gaps in the sequence numbers of the received packets are used to detect lost packets.

Let the segment size be $n$ bits. Then, each $w$-bit subsequence of encoded data corresponds to $\nu$ segments. Let a single packet be lost in the network. The payloads of the packets corresponding a message are reassembled at the receiver, with an $n$-bit gap corresponding to the segment that was in the payload of the lost packet. Let the lost segment be $\mathbf{r}_i$. Each inversion is based on a $w$-bit received subsequence, i.e., on $\nu$ $n$-bit received segments. In the absence of any random errors, the estimated encoded sequence $\hat{\mathbf{v}}$ is the same as the received sequence $\mathbf{r}$, and the lost segment is $\hat{\mathbf{v}}_i$. Each $w$-bit received subsequence is mapped into its corresponding $w$-bit information subsequence by using (4.11). The subsequences that include the lost segment cannot be inverted, causing a *inversion skip.*

For the local inversion to tolerate a single lost packet, the $w$-bit information subsequences $\hat{\mathbf{u}}_{[i-nl,i-1]}$ and $\hat{\mathbf{u}}_{[i+1,i+nl]}$, corresponding to the $w$-bit estimated encoded subsequences $\hat{\mathbf{v}}_{[i-\nu,i-1]}$ and $\hat{\mathbf{v}}_{[i+1,i+\nu]}$, respectively, should either overlap or be adjacent and contiguous to each other. There are $\nu$ inversion skips between the estimated information subsequences $\hat{\mathbf{u}}_{[i-nl,i-1]}$ and $\hat{\mathbf{u}}_{[i+1,i+nl]}$. Hence, the subsequence $\hat{\mathbf{u}}_{[i+1,i+nl]}$ is shifted $(\nu+1)k$ bits with respect to the subsequence $\hat{\mathbf{u}}_{[i-nl,i-1]}$. In order to successfully recover the lost segment, we must have

$$w - (\nu + 1)k \geq 0. \tag{11.1}$$

Substituting $w = n\nu = n(K - k)$, $K = \nu + k$, and $k = n - 1$, we get

$$K - 2k \geq 0. \tag{11.2}$$

This implies that the input constraint length should be at least twice the numerator of the coding rate in order to recover a single lost segment. A necessary condition for recovering a single lost segment is that the $\nu$ segments immediately preceding and succeeding the lost segment should be received correctly. This is also known as the *guard space* requirement.

**Example 11.2** Suppose that the encoded sequence $\mathbf{v}_{[0,6]}$ = 111010 100 011001 101110 in Example 11.1 is segmented into a series of 3-bit segments 111 010 100 011 001 101 110, and that these segments are appropriately numbered, encapsulated, and transmitted over the network. Next, assume that there are no random errors during transmission due to channel noise and that a packet is lost during transmission due to buffer overflow at an intermediate switching station. Let the packet carrying the fourth 3-bit segment 011 be lost. The received message is now 111 010 100 (- - -) 001 101 110, where the "oldest" segment is 010. The position of the lost segment within the sequence is known and is indicated with "(- - -)". Using (4.11), the 6-bit encoded subsequences $\mathbf{v}_{[0,1]}$ = 111010, $\mathbf{v}_{[1,2]}$ = 010100, $\mathbf{v}_{[4,5]}$ = 001101, and $\mathbf{v}_{[5,6]}$ = 101110 are inverted to get the 6-bit information subsequences $\mathbf{u}_{[0,2]}$ = 000110, $\mathbf{u}_{[1,3]}$ = 011010, $\mathbf{u}_{[4,6]}$ = 110111, and $\mathbf{u}_{[5,7]}$ = 011100, respectively. The missing segment causes two inversion skips, corresponding to the encoded blocks $\mathbf{v}_{[2,3]}$ = 100(- - -) and $\mathbf{v}_{[3,4]}$ = (- - -) 001. However, the decoded blocks $\mathbf{u}_{[1,3]}$ and $\mathbf{u}_{[4,6]}$ are contiguous and the original message is entirely recovered.

The inversion step can be implemented using a look-up table to store the inverse $[\mathbf{G}]_{6\times 6}^{-1}$. However, the table size grows quadratically with the one-to-one mapping length $w$, and becomes prohibitively large for coding rates and lengths of practical interest. The hardware complexity of the decoder can be reduced by implementing the multiplication in (4.11) using two shift registers of length $w$, as shown in Figure 11.2 for the rate two-thirds, $K = 4$ encoder of Example 11.2. Additional control hardware is also required which is not shown in this

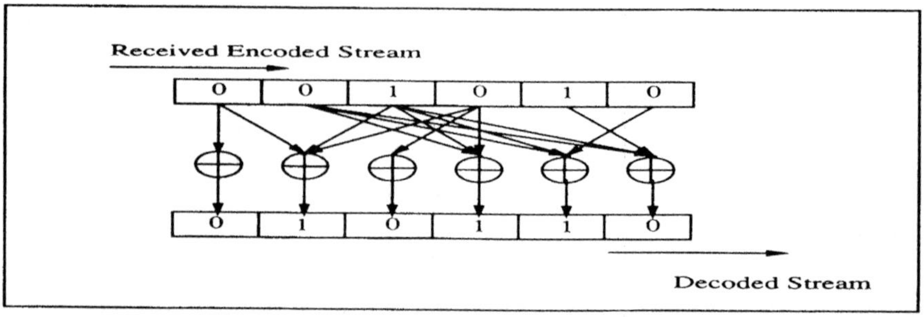


**Figure 11.2** Decoder for a rate two-thirds, $K = 4$ convolutional encoder.

figure for simplicity. The design of the complete lost packet recovery hardware is the subject of ongoing research.

### 11.3.3 Recovery of Multiple Lost Packets

The single lost packet recovery scheme in Example 11.2 would not work if two or more consecutive packets were lost in transmission. If two consecutive packets were lost, there would be three inversion skips, and the 6-bit estimated information subsequences immediately before and after the skips would no longer be consecutive. We can generalize the method outlined above for recovering multiple consecutive lost packets. Let the $r$ consecutive $n$-bit segments $\hat{\mathbf{v}}_i$, $\hat{\mathbf{v}}_{i+1}$, ..., $\hat{\mathbf{v}}_{i+r-1}$. be lost. For the local inversion to tolerate $r$ lost packets, the $w$-bit information subsequences $\hat{\mathbf{u}}_{[i-nl,i-1]}$ and $\hat{\mathbf{u}}_{[i+r,i+r+nl-1]}$, corresponding to the $w$-bit estimated encoded subsequences $\hat{\mathbf{v}}_{[i-\nu,i-1]}$ and $\hat{\mathbf{v}}_{[i+r,i+r+\nu-1]}$, respectively, should either overlap or be adjacent and contiguous to each other. This condition can be expressed as follows:

$$w - (\nu + r)k \geq 0. \tag{11.3}$$

Substituting $w = n\nu = n(K - k)$, $K = \nu + k$, and $k = n - 1$, we get

$$K - (r + 1)k \geq 0. \tag{11.4}$$

The relation in (11.4) imposes a restriction on the input constraint length based on a given code rate and the number of lost segments that we wish to reconstruct.

**Example 11.3** From (11.4), it is clear that if $r \geq 2$, the rate two-thirds, $K = 4$ encoder of Example 11.1 will not work. The code rate and/or the input constraint length need to be changed. If the code rate is kept constant at two-thirds, and $r = 4$ consecutive lost segments are to be recovered, the input constraint length $K$ has to be at least 10, and correspondingly, $w \geq 24$. Thus, recovery of four consecutive lost segments using a rate two-thirds encoder requires access to a matrix of size $24 \times 24$, which is considerably larger than the matrix required for recovering a single lost segment using an encoder of the same coding rate. The code rate can also be reduced to reduce the one-to-one mapping length. For instance, if the code rate is reduced to 1/3, then an encoder with $K = 5$ will allow recovery of four consecutive lost segments.

The examples considered here are illustrative of the design trade-offs necessary for selecting an appropriate encoder. In practice, much larger code rates and input constraint lengths are required.

### 11.3.4 Application to ATM Networks

A typical SAR sublayer segment size recommended for connectionless service is 44 bytes (352 bits). This segment is encapsulated with a header and a trailer of 2 bytes each, thus making up the 48 byte payload of an ATM cell. Suppose that we want to recover one lost segment. Choosing a segment size of 352 bits makes $n = 352$. We will consider various basic coding rates $k'/n'$ (ranging from 1/2 to 31/32), where $k = mk'$ and $n = mn'$ for some positive integer $m$. Having fixed the denominator $n$, the numerator $k$ can be calculated for each of the code rates. Next, we need to select the input constraint length. From (11.2), the minimum value that the input constraint length $K$ can take is $2k$. Here we restrict ourselves to the minimum possible $K$. The one-to-one mapping window length $w$ is then calculated from (4.8). The basic code rates and the corresponding one-to-one mapping length for recovering a single lost ATM cell are shown in Table 11.2. It follows that low coding redundancies require a larger $w$. If we choose to keep the code rate fixed at one-half, we can recover multiple lost packets by increasing $K$ appropriately. A larger $K$ also implies a larger $w$. Table 11.2 shows the one-to-one mapping length necessary to recover up to 5 lost packets using rate one-half convolutional coding. The approach

| | Basic Rate $k'/n'$ ($r = 1$) | | | | | No. Lost Cells $r$ ($k'/n' = 1/2$) | | | | |
|---|---|---|---|---|---|---|---|---|---|---|
| | 1/2 | 3/4 | 7/8 | 15/16 | 31/32 | 1 | 2 | 3 | 4 | 5 |
| $k$ | 176 | 264 | 308 | 330 | 341 | 176 | 176 | 176 | 176 | 176 |
| $n$ | 352 | 352 | 352 | 352 | 352 | 352 | 352 | 352 | 352 | 352 |
| $K$ | 352 | 528 | 616 | 660 | 682 | 352 | 528 | 704 | 880 | 1056 |
| $w$ | 352 | 1056 | 2464 | 5280 | 10912 | 352 | 704 | 1056 | 1408 | 1760 |

**Table 11.2** Parameters for convolutional lost cell recovery in ATM networks.

presented in this section makes convolutional coding a possible alternative to the traditional block coding techniques for recovery of consecutive lost cells.

### 11.3.5 Performance

Let the generation of ATM cells be modeled as a data source, and let the network be represented as a single server with a finite buffer. A cell arriving at the server is discarded if the buffer is full, resulting in cell loss. Assume that no random errors occur in the network. Time is slotted and each cell requires exactly one slot to be transmitted. The buffer size of the server is denoted by $B$. Further, assume that the number of packets arriving in time slots 1, 2, ... are independent, Poisson distributed random variables with rates $\lambda$. The effective rate for the coded cases is $\lambda_{eff} = \lambda(\frac{n}{k})$. Each cell finds the buffer full with a probability $p$ independently of other cells. This probability is the rejection probability $P_{loss}(\lambda)$ (or $P_{loss}(\lambda_{eff})$) of the single server queue modeled as a finite-state, discrete time Markov chain in which the state is the number of cells in the queue just before the beginning of a slot and no retransmissions occur. Examples of packet loss probability computations are given in Appendix 11.5 of this chapter.

The probability of $i$ lost cells in an encoded user PDU of length N cells is

$$P_i = \binom{N}{i} p^i (1-p)^{N-i}. \tag{11.5}$$

If $r = 1$, then the expected number of lost cells per PDU after decoding is

$$E_N = \sum_{i=2}^{N} i P_i. \tag{11.6}$$

The packet loss rate after decoding is

$$P_{dec}(\lambda_{eff}) = E_N/N. \tag{11.7}$$

The performance of the scheme is measured by the *loss ratio* $G$, defined as [412]

$$G = P_{dec}(\lambda_{eff})/P_{loss}(\lambda). \tag{11.8}$$

It is the reduction in cell loss probability due to coding. The loss ratio should satisfy $G < 1$ in order for coding to be useful. In Table 11.3, the loss ratios attainable for a range of code rates and packet arrival rate $\lambda$ are shown. The loss ratio improves with coding efficiency because fewer additional cells result in a lower $P_{loss}(\lambda_{eff})$. Buffer size is $B = 20$ and the user PDU length is $N = 50$. Some of the loss ratios are not shown in Table 11.3 because coding did not offer any advantage in these cases.

| | | Loss Ratio | | | |
|---|---|---|---|---|---|
| $\lambda$ | $P_{loss}(\lambda)$ | 63/64 | 31/32 | 15/16 | 7/8 |
| 0.7 | 3.4060e-07 | 1.6115e-04 | 2.8867e-04 | 7.3338e-04 | 9.5395e-02 |
| 0.8 | 2.5500e-05 | 3.8500e-03 | 1.1613e-02 | 1.0801e-01 | |
| 0.9 | 1.3547e-03 | 1.7191e-01 | 4.3657e-01 | | |

**Table 11.3** Loss ratio $G$ for various coding and cell arrival rates (B=20, N=50).

### 11.3.6 Encoder Selection

Given the code rate and the input constraint length, it is necessary to identify desirable properties of encoders that can be used for lost packet recovery. Under the assumption of no random errors, the distance parameters of the encoder are not important. This means that encoders with either small or large distance parameters would work equally well as long as they are locally invertible. However, searching of locally invertible encoders is highly computationally intensive, even if examining the distance structure is not necessary. This is because typical input constraint lengths and rates desired may be quite different from the convolutional encoders used for random error correction. Thus, listings of convolutional encoders available in the literature are not of much use. An algorithmic approach for finding long locally invertible encoders is one of the topics that needs to be researched further.

## 11.4 PERFORMANCE IMPROVEMENT TECHNIQUES

The performance of a lost packet recovery scheme can be further improved by addressing certain practical issues. We will discuss some of them here.

### 11.4.1 Interleaving and Buffer Management

Packets are typically lost in bursts which results in an increase in the number of consecutive lost packets that need to be recovered. This number can be effectively reduced by interleaving the packet stream going out of the transmitting end, resulting in a reduction in the required guard space around the lost packet and making the scheme more economical. Buffer management at intermediate switches can be used to spread the effect of packet loss across multiple virtual paths and virtual circuits. The interleaving can be deterministic (assigning every $M$th packet to a different stream) or random (multiplexing packets arriving from many sources) [412]. The guard space requirement must also be closely matched with the duty cycle of the bursts of lost packets. For minimizing coding overhead, a high basic code rate is preferable. The average "quiet" length, i.e., the number of consecutive packets received correctly between any two bursts of lost packets must be large enough to accommodate the necessary guard space requirement. Excessively high level of interleaving decreases this "quiet" length for each packet stream and may result in insufficient guard space around the lost packet.

### 11.4.2 Random Errors As Lost Packets

The random noise on the transmission channel can introduce errors in the packets. Errors in the header of the ATM cells can be detected since a CRC checksum is included as part of the header. These errors usually result in a retransmission request for the entire PDU. We can treat such errors using the scheme described above by simply declaring the erroneous packet as a lost packet. The convolutional coding used for lost packet recovery can also be applied to correct random errors if they occur. The combined lost packet recovery and error correcting capability of convolutional decoding approach described in this chapter is also a topic of future research.

## 11.5 EXERCISES

1. Consider the (5, 3, 2) encoder with generator sequences $G^{(1)} = 101\ 1101$, $G^{(2)} = 011\ 0111$, $G^{(3)} = 110\ 1001$, $G^{(4)} = 010\ 1010$, and $G^{(5)} = 111\ 1101$. Determine if this encoder is locally invertible by finding its reduced encoding matrix and checking its invertibility. What is the input constraint length $K$ of this encoder? How many consecutive 5-bit lost segments can be recovered using this encoder?

2. Repeat Exercise 1 for the (3, 1, 4) encoder with generator sequences $G^{(1)} = 10101$, $G^{(2)} = 01101$, and $G^{(3)} = 11111$.

3. Compute guard space requirement on each side of the lost segments for the encoder of Exercise 2.

4. By trial-and-error methods, construct a $(3, 2, m)$ locally invertible encoder that can recover two consecutive 3-bit lost segments.

5. Write a computer program to perform lost packet recovery using the encoder of Exercise 2. Your program should incorporate the following: encoding of a user PDU of a specified length, segmentation of the encoded sequence and assignment of sequence numbers to the segments, loss of some segments according a network model of your choice, and recovery of the lost segments. Your program should check whether the guard space requirements are met, and should flag if lost segments cannot be recovered due to insufficient guard space.

# APPENDIX A

# Packet Loss Probability

The calculation of packet loss probability depends on the network model. Two scenarios for are described here. Interested readers may refer to [246, 203, 45, 342] and [9, 430, 43, 184].

## *Output Queueing*

We summarize the computation of probability of packet loss in an $N \times N$ high-speed switch that uses output queueing to buffer packets contending for the same outgoing link. Output queueing achieves optimal throughput/delay performance. For further details, see [203, 246].

Let the switching fabric be an $N \times N$ switch. Each output line has a separate $b$ packet FIFO queue at the output of the switching fabric. Let $A$ be a random variable denoting the the number of packet arrivals destined for a particular (tagged) output in a given time slot. The probability of $k$ arrivals in a time slot is given by

$$a_k = Pr[A = k] = \binom{N}{k} (p/N)^k (1 - p/N)^{N-k}, \; k = 0, \, 1, \, \ldots, \, N. \quad \text{(A.1)}$$

For the limiting case of $N = \infty$, we get

$$a_k = Pr[A = k] = \frac{p^k e^{-p}}{k!}, \; k = 0, \, 1, \, \ldots, \, N = \infty. \quad \text{(A.2)}$$

Let $Q_m$ be the queue size at the end of the $m$th time slot, and let $A_m$ be the number of packet arrivals during the $m$th time slot. Then

$$Q_m = \min \{\max (0, \, \mathrm{Q_{m-1} + A_m - 1}), \, \mathrm{b}\}. \quad \text{(A.3)}$$

One of the arriving packet is immediately transmitted during the $m$th time slot if $Q_{m-1} = 0$ and $A_m > 0$. Clearly, if the queue is empty at the end of the previous time slot, one of the arriving packet flows through the switch without delay. For $N = \infty$ and $b = \infty$, the queue size $Q_m$ can be modeled by a $M/D/1$ queue. For finite $N$ and $b$, the queue size can be modeled by a finite-state, discrete-time Markov chain with state transition probabilities $P_{ij} = Pr[Q_m = j \mid Q_{m-1} = i]$ given by

$$P_{ij} = \begin{cases} a_0 + a_1 & i = 0, \, j = 0 \\ a_0 & 1 \le i \le b, \, j = i - 1 \\ a_{j-i+1} & 1 \le j \le b - 1, \, 0 \le i \le j \\ \sum_{m=j-i+1}^{N} a_m & j = b, \, 0 \le i \le j \\ 0 & \text{otherwise,} \end{cases} \quad \text{(A.4)}$$

where $a_k$ is given by (A.1) and (A.2) for $N < \infty$ and $N = \infty$, respectively.

The steady-state queue size, obtained directly from the Markov chain balance equations, is

$$q_1 = Pr[Q = 1] = \frac{1 - a_0 - a_1}{a_0} \cdot q_0 \quad \text{(A.5)}$$

$$\vdots$$

$$q_n = Pr[Q = n] = \frac{1 - a_1}{a_0} \cdot q_{n-1} - \sum_{k=2}^{n} \frac{a_k}{a_0} \cdot q_{n-k}, \; 2 \leq n \leq b, \tag{A.6}$$

where

$$q_0 = Pr[Q = 0] = \frac{1}{1 + \sum_{n=1}^{b} q_n / q_0}. \tag{A.7}$$

A packet will not be transmitted on the tagged output line during the $m$th time slot if and only if $Q_{m-1} = 0$ and $A_m = 0$. Let $\rho_0$ be the normalized switch throughput. Then,

$$\rho_0 = 1 - q_0 a_0. \tag{A.8}$$

A packet will be lost if it finds the output queue already containing $b$ packets. The packet success probability is obtained by dividing the utilization of the output line $\rho_0$ by the arrival rate $p$. Then,

$$Pr[\text{packetloss}] = 1 - \frac{\rho_0}{\text{p}}. \tag{A.9}$$

## *Tandem M/M/1/B Queues*

Consider $L$ $M/M/1/B$ queues in tandem. The interarrival times and service times are exponentially distributed. There is a single server, and each queue has a finite maximum size equal to $B$. The buffer at the source is assumed to be infinite. Let

$$\rho_i = \frac{\lambda_i}{\mu_t} \quad i = 1, 2, \ldots, L, \tag{A.10}$$

where $\mu_t$ is constant since each network node is assumed to have identical parameters. Then, the average queue length for the $i$th queue is given by

$$E[L_i] = \begin{cases} \frac{\rho_i}{(1-\rho_i)} & i = 1 \\ \frac{\rho_i}{(1-\rho_i)} - \frac{(B+1)\rho_i^{B+1}}{1-\rho_i^{B+1}} & i = 2, \ldots, L. \end{cases} \tag{A.11}$$

The probability $q_i$ that the buffer at node $i$ is full can also be determined. It is given by [9]

$$q_i = \frac{(1 - \rho_i)\rho^{B+1}}{1 - \rho_i^{B+1}} \quad i = 2, \ldots, L. \tag{A.12}$$

We can obtain the solution for this equation iteratively by starting with the destination node, since

$$\lambda_L = \frac{\lambda_{VC}}{1 - q_L}, \tag{A.13}$$

where $\lambda_{VC}$ is the average arrival rate. The boundary conditions are that for $\lambda_{VC} = 0$, we have $q_L = 0$, and a numerical procedure can be used to get $q_L$ and $\lambda_L$. We then make use of the fact that

$$\lambda_i = \frac{\lambda_{i+1}}{1 - q_i}, \quad i = 2, \ldots, L - 1 \tag{A.14}$$

to solve for $q_i$ for $i = L - 1, \ldots, 2$. Finally, with an infinite buffer at the source, we have $\lambda_1 = \lambda_2$.

# A

# BASIC CONCEPTS IN ALGEBRA

This appendix briefly describes some of the basic structures in algebra and is intended to serve as a quick reference for terms and concepts used in the book. Interested readers may referred to [185, 95, 211, 279, 325, 383].

## A.1 SETS AND OPERATIONS

### A.1.1 Sets

A *set* $S$ is a collection of objects $s,\ s_1,\ s_2,\ \ldots$, which are called the *elements* of this set. We say that $s \in S$ if $s$ is an element of the set $S$. Familiar examples of sets are the set of integers $\mathcal{Z}$, the set of rational numbers $\mathcal{Q}$, and the set of real numbers $\mathcal{R}$. An example of a finite set is $\{0,\ 1\}$.

A set with no elements in it is called the *empty* set and is denoted as $\emptyset$ or as $\{\ \}$. When two sets have exactly the same elements, they are said to be *equal.* The *union* of two sets $S$ and $T$ is the set $S \cup T = \{x : x \in S \textit{ or } x \in T(\text{or both})\}$, and their *intersection* is the set $S \cap T = \{x : x \in S \textit{ and } x \in T\}$. The *difference* of two set $S$ and $T$ is the set $S \setminus T = \{s : s \in S \text{ and } s \notin T\}$.

The *Cartesian product* (or simply the *product*) of two sets $S_1$ and $S_2$ is denoted by $S_1 \times S_2$ and is the set $S_1 \times S_2 = \{(s_1,\ s_2) : s_1 \in S_1,\ s_2 \in S_2\}$. That is, it is the set of all ordered pairs in which the first component is a member of $S_1$ and the second component is a member of $S_2$. The product of a set $S$ with itself is the set $S \times S = \{(s,\ t) : s,\ t \in S\}$ and is also called the set of ordered pairs of elements of $S$.

## A.1.2 Binary Operations on a Set

An *operation* on the elements of a set is a method of combining two or more elements to obtain another element. An operation involving two elements is called a *binary operation.* Addition and multiplication as defined on the set of integers or real numbers are examples of binary operations. Let '$*$' and '@' be two binary operations defined on the set $S$. Let $s_1, s_2, s_3 \in S$. The following properties are of interest.

P1. *Closure*: For all $s_1, s_2 \in S$, $s_1 * s_2 \in S$, i.e., the outcome of performing the operation is an element which is also a member of $S$.

P2. *Associativity*: $s_1 * (s_2 * s_3) = (s_1 * s_2) * s_3$, i.e., the order in which the elements are paired to perform the operation does not matter.

P3. *Commutativity*: $s_1 * s_2 = s_2 * s_1$, i.e., the order in which the operation is performed on the elements does not matter.

P4. *Distributivity of @ over* $*$: $s_1@(s_2 * s_3) = s_1@s_2 * s_1@s_3$.

P5. *Identity element* ($e$): $e * s = s * e = s$, i.e., an element which has the property that it leaves the other element unchanged in an operation.

P6. *Inverse of an element* ($s^{-1}$): $s * s^{-1} = e$, $s^{-1} * s = e$. More precisely, an element $s^{-1}$ is called a *right inverse* of $s$ if $s * s^{-1} = e$, and it is called a *left inverse* if $s^{-1} * s = e$. If the right and the left inverses of an element are the same, they are simply called the *inverse*. Note that if the operation is commutative, the right and the left inverses are automatically the same.

## A.1.3 Subsets

A set $T$ is said to be a *subset* of $S$ if all the elements of $T$ are also the elements of $S$. This is denoted as $T \subseteq S$. For example, the set of even integers is a subset of $\mathcal{Z}$. If there is at least one element of $S$ that is not an element of $T$, then $T$ is called a *proper* subset of $S$ ($T \subset S$). The empty set is a subset of all sets. Two set $S$ and $T$ are equal if and only if $S \subseteq T$ and $T \subseteq S$. The set of all subsets of a set $S$ is called the *power* set of $S$ and is denoted as $\mathcal{P}(S)$.

## A.1.4 Relations and Partitions

A *relation* from a set $S$ to a set $T$ is a subset $R$ of $S \times T$. If $(s,t) \in R$, then $s$ and $t$ are "related" to each other. This is also written as $s\ R\ t$. A relation from $S$ to $S$ is called a *relation on* $S$.

A special kind of relation is the *equivalence* relation, denoted by '$\sim$'. Given an equivalence relation $E$, an ordered pair $(a,\ b) \in E$ if $a \sim b$. An equivalence relation $E$ satisfies the following properties: 1) *Reflexivity*, $(a,\ a) \in E$, i.e., $a \sim a$; 2) *Symmetry*, if $(a,\ b) \in E$, then $(b,\ a) \in E$; i.e., if $a \sim b$, then $b \sim a$; 3) *Transitivity*, if $(a,\ b) \in E$, and $(b,\ c) \in E$, then $(a,\ c) \in E$, i.e., if $a \sim b$ and $b \sim c$, then $a \sim c$. Let $S$ be a non-empty set and let $S_I$ be any collection of non-empty subsets $S_i$, $i \in I$ of $S$ such that the union of the sets in $S_I$ is $S$, and each distinct pair of sets in $S_I$ has empty intersection. Then $S_I$ (where $I$ is an indexing set) is called a *partition* of $S$.

For a given element $s \in S$ and an equivalence relation $R$ defined on $S$, the set of all elements of $S$ that are equivalent to $s$ is called an *equivalence class*, and is denoted by $[s] = \{t \in S : (s,\ t) \in R\}$. Two equivalence classes either coincide or are disjoint. If all the elements of a set $S$ are divided into equivalence classes, the result is a set of disjoint subsets of $S$ and hence a *partition* of $S$. Thus, an equivalence relation induces on a set a partition of disjoint subsets.

## A.1.5 Functions

A *function* (or a *mapping*) from a set $S$ to a set $T$ is a rule for assigning to each element of $S$ exactly one element of $T$. More precisely, a function $f$ from $S$ to $T$ is a relation $R$ from $S$ to $T$ such that for each $s \in S$ there exists exactly one $t \in T$ for which $(s,\ t) \in f$. The set $S$ is called the *domain* of $f$, and the set $T$ is called the *codomain* of $f$. The set $f(S) = \{t \in T : t = f(s)$ for some $s \in S\}$ is called the image of $f$. The function $f$ is called *one-to-one* (1-1), or *injective*, if whenever $f(s_1) = f(s_2)$, then $s_1 = s_2$. The function $f$ is called *onto*, or *surjective*, if for all $t \in T$ there exists $s \in S$ such that $f(s) = t$. If the function is both 1-1 and onto, it is called a *1-1 correspondence* or a *bijection*. A function $f \subseteq S \times T$ from $S$ to $T$ is denoted by $f : S \to T$.

We define a binary operation $*$ as a mapping from $* : S \times S$ into $S$. The operation $*(s_1,\ s_2) = s$ is usually written as $s_1 * s_2 = s$. For the sets of integers, rational numbers, and real numbers, addition and multiplication are binary operation and can also be thought of as functions from $\mathcal{Z} \times \mathcal{Z}$ to $\mathcal{Z}$.

# A.2 GROUPS, RINGS, FIELDS, AND VECTOR SPACES

## A.2.1 Groups

A set $G$ under a binary operation denoted by '$*$' is called a *group*, written as $(G, *)$, if the following properties are satisfied: closure (P1), associativity (P2), identity (P5), and inverse (P6). It is called *abelian* (or *commutative*) if it also satisfies commutativity (P3). It can be shown that the identity and the inverse element of a group are unique. The number of elements in a group $G$ is called its *order* and is denoted as $|G|$. A group of finite order is called a *finite* group.

**Example A.1** A group we commonly encounter is $(\mathcal{Z}, +)$, where $+$ is the ordinary addition defined on integers. In $(\mathcal{Z}, +)$, 0 is the identity element, and negative of an integer is its inverse. In fact, $(\mathcal{Z}, +)$ is an abelian group. $(\{0, 1\}, +_2)$ is a group if $+_2$ denotes addition modulo 2. It is a finite group.

### *Subgroups*

Let $(G, *)$ be a group and let $H$ be a non-empty subset of $G$. Then $H$ is a *subgroup* of $G$ if $H$ is a group under the restriction of $*$ to $H$. The subgroup of $(G, *)$ consisting of all powers of the element $g$ of $G$ is called the subgroup *generated* by $g$ and is denoted by $\langle g \rangle$. A subgroup $(H, *)$ of the group $(G, *)$ is called a *normal* subgroup if $g * H * g^{-1} \in H$ for all $g \in G$. Every subgroup of an abelian group is normal.

### *Cosets*

Let $(H, *)$ be a subgroup of $(G, *)$. For a fixed $g \in G$, the set $g * H = \{g * h \mid h \in H\}$ is called the *left coset* of $(H, *)$ in $(G, *)$. Similarly, the set $H * g = \{h * g \mid h \in H\}$ is called the *right coset* of $(H, *)$ in $(G, *)$. If the left coset $g * H$ of $(H, *)$ is equal to the right coset $H * g$ for all $g \in G$, then $(H, *)$ is called a *normal* subgroup of $(G, *)$. The left and the right cosets of $(H, *)$ are the same if $(G, *)$ is abelian. Hence, every subgroup of an abelian group is normal.

Two cosets of $(H, *)$ in $(G, *)$ either coincide or are disjoint. Thus, cosets of $(H, *)$ induce a partition on $(G, *)$. A subgroup $(H, *)$ of $(G, *)$ is its own coset in $(G, *)$. The only coset of $(H, *)$ in $(G, *)$ that is a subgroup of $(G, *)$ is $H$ itself since it is the only coset containing the identity element.

For a finite group $(G, *)$, the number of distinct left (or right) cosets of $(H, *)$ in $(G, *)$, denoted as $[G : H]$, is the same and is called the *index* of $(H, *)$ in $(G, *)$. The mapping $H \to g * H : h \to g * h$, where $g \in G$, is 1-1 and onto. Hence, there is a 1-1 correspondence between a subgroup and any of its left cosets, and the cosets of $(H, *)$ in $(G, *)$ partition $G$ into subsets all containing the same number of elements as $H$. Therefore, the order of a subgroup divides the order of the group. That is, for a finite group $(G, *)$ and its subgroup $(H, *)$, we have $|G| = [G : H] \cdot |H|$. This is Lagrange's Theorem.

## *Quotient (Factor) Groups*

For a normal subgroup $(H, *)$ of the group $(G, *)$, the set of left cosets of $(H, *)$ in $(G, *)$ forms a group under the operation $(g_1 * H) \cdot (g_2 * H) = (g_1 * g_2) * H$. This group is called the *factor group* (or the *quotient group*) of $G$ modulo $H$ and is denoted by $G/H$. In the factor group, the normal subgroup acts as the identity element, since $(g * H) \cdot (h * H) = (g * h) * H = g * H$ for $g \in G$ and $h \in H$.

# A.2.2 Rings

A set $R$ under two binary operations $+$ (called addition) and $\cdot$ (called multiplication) is called a *ring*, denoted as $(R, +, \cdot)$, if the following properties are satisfied: $(R, +)$ is a commutative group, closure of $\cdot$ (P1), associativity of $\cdot$ (P2), and distributivity of $\cdot$ over $+$ (P4). Note that multiplication is not as "strong" as addition in a ring. Addition satisfies properties P1, P2, P3, P5, and P6, while multiplication only satisfies P1 and P2. The ring takes on additional properties as multiplication becomes "stronger". If the set $R$ has a multiplicative identity (P5), then $(R, +, \cdot)$ is called a *ring with identity*. If $\cdot$ is commutative (P3), then $(R, +, \cdot)$ is called a *commutative* ring. If in a commutative ring with identity $R$, $r_1 \cdot r_2 = 0$ implies $r_1 = 0$ or $r_2 = 0$, for all $r_1, r_2 \in R$, then $(R, +, \cdot)$ is called an *integral domain*. If $(R \setminus \{0\}, \cdot)$ forms a group, then $(R, +, \cdot)$ is called a *division ring*. This means that the ring has a multiplicative identity (different from the additive identity 0), and all nonzero elements of the ring have multiplicative inverses (i.e., it is possible to *divide* in the ring).

**Example A.2** The set of integers $\mathcal{Z}$ with addition and multiplication operations is a ring since multiplication is associative and distributive over addition. It is easy to verify that $(\mathcal{Z}, +, \cdot)$ is in fact an integral domain. The rational numbers $\mathcal{Q}$ and the real numbers $\mathcal{R}$ are also rings under addition and multiplication. The set $\{0, 1\}$ is a ring under addition and multiplication modulo 2 ($+_2$ and $\cdot_2$, respectively).

Let $(R, +, \cdot)$ be a ring. If there exists a positive integer $n$ such that $nr = r + r + \cdots + r = 0$ ($n$ summands $r$) for all $r \in R$, then the least such $n$ is called the *characteristic* of $(R, +, \cdot)$. If no such positive integer exists, then the characteristic of $(R, +, \cdot)$ is 0.

## *Subrings*

Let $(R, +, \cdot)$ be a ring and let $S$ be a subset of $R$. Then $(S, +, \cdot)$ is called a *subring* of $(R, +, \cdot)$ if it is a ring under the same operations as defined on $R$.

## *Ideals*

Let $(R, +, \cdot)$ be a ring and let $(J, +)$ be a subgroup of $(R, +)$. Note that $(J, +)$ is a normal subgroup since $(R, +)$ is an abelian group. Then $J$ is called an *ideal* of $(R, +, \cdot)$ if it is a subring of $(R, +, \cdot)$ and, for all $j \in J$ and $r \in R$, we have $j \cdot r \in J$ and $r \cdot j \in J$.

If $(R, +, \cdot)$ is a commutative ring with identity, then the smallest ideal containing a given element $s \in R$ is the ideal $(r) = \{r \cdot s : r \in R\}$. If there exists a $j \in J$ such that $J = j \cdot R$, then $J$ is said to be a *principal* (left) ideal. It is also called the principal ideal *generated* by $j$ and denoted by $(j)$. If $(R, +, \cdot)$ is an integral domain in which every ideal is principal, then $(R, +, \cdot)$ is called a *principal ideal domain.* For example, $(\mathcal{Z}, +, \cdot)$ is a principal ideal domain.

Let $(R, +, \cdot)$ be an integral domain. Any $r \in R$ which has a multiplicative inverse in $R$ is called a *unit.* Any $r, s \in R$ such that $r \cdot s = 1$ are called *associates.* A nonzero nonunit $p \in R$ such that if $p|r \cdot s$ implies $p|r$ or $p|s$ is called a *prime.* A nonzero nonunit $p \in R$ such that if $p = r \cdot s$ implies $r$ or $s$ is a unit is called *irreducible.*

## *Quotient (Residue Class) Rings*

An ideal $J$ is a normal subgroup of the additive group $(R, +)$ of the ring $(R, +, \cdot)$. Hence, the ideal induces a partition of $R$ into disjoint cosets called *residue classes* modulo $J$. The residue class of the element $r$ is denoted by $[r] = r + J$, since it contains elements of $R$ that are of the form $r + j$ for some $j \in J$. We know from the previous section that $R/J$ is a quotient group since $J$ is a normal subgroup of $(R, +)$. Define two operation $(r_1 + J) + (r_2 + J) = (r_1 + r_2) + J$, and $(r_1 + J) \cdot (r_2 + J) = (r_1 \cdot r_2) + J$. The set of residue classes of a ring $(R, +, \cdot)$ modulo an ideal $J$ forms a ring with respect to these two operations. This ring is called the *residue class ring* (or *factor ring* or *quotient ring*) of $R$ modulo $J$ and is denoted by $R/J$.

### A.2.3 Fields

A set $F$ under two binary operations $+$ (called addition) and $\cdot$ (called multiplication) is called a *field*, and denoted as $(F, +, \cdot)$, if the following properties are satisfied: $(F, +)$ is a commutative group, $(F \setminus \{0\}, \cdot)$ is a commutative group (where 0 is the additive identity), and $\cdot$ is distributive over $+$ (P4). Note that a field is a commutative division ring.

Let $(F, +, \cdot)$ be a field. If there exists a positive integer $n$ such that $nf = f + f + \cdots + f = 0$ ($n$ summands $f$) for all $f \in F$, then the least such $n$ is called the *characteristic* of $(F, +, \cdot)$. If no such positive integer exists, then the characteristic of $(F, +, \cdot)$ is 0. A field with a finite characteristic is called a *finite* field.

**Example A.3** The rings $(\mathcal{Q}, +, \cdot)$ and $(\mathcal{R}, +, \cdot)$ are fields. The ring of integers $(\mathcal{Z}, +, \cdot)$ is not a field since nonzero integers are not a group under multiplication. The set $\{0, 1\}$ is a field under addition and multiplication modulo 2. It is a finite field. Another example of a finite field is $\mathcal{Z}/p\mathcal{Z}$, the field of integers modulo a prime $p$.

## *Subfields*

Let $(F, +, \cdot)$ be a field. A subset $D$ of $F$ is a *subfield* of $(F, +, \cdot)$ if it is a field under the same operations as defined on $F$. The intersection of all subfields of $(F, +, \cdot)$ is a unique smallest subfield $(P, +, \cdot)$ called the *prime subfield* of

$(F, +, \cdot)$. If $(F, +, \cdot)$ is a subfield of $(K, +, \cdot)$, then $(K, +, \cdot)$ is called an *extension field* of $(F, +, \cdot)$.

### *Quotient Fields*

An integral domain $(D, +, \cdot)$ is contained in a field $(F, +, \cdot)$ such that for every $f \in F$, $f = g \cdot h^{-1}$ with $g, h \in D$. Such a field is a *field of quotients* of $D$. For example, the integral domain $(\mathcal{Z}, +, \cdot)$ is a subring of the field $(\mathcal{Q}, +, \cdot)$, and every rational number can be written in the form $z_1 \cdot z_2^{-1}$ with $z_1, z_2 \in \mathcal{Z}$.

## A.2.4 Vector Spaces

Let $(F, +, \cdot)$ be a field and let its elements be called *scalars*. Let $V$ be a set whose elements are called *vectors*. Define two operations: *vector addition* $\bar{+} : V \times V \to V$ such that for $v_1, v_2 \in V$, $v_1 \bar{+} v_2 \in V$; and *scalar multiplication* $\bar{\cdot} : F \times V \to V$ such that for $f \in F$, $v \in V$, $f \bar{\cdot} v \in V$. Then $V$ is called a *vector space* over $F$ if following properties are satisfied:

(i) $(V, \bar{+})$ is an abelian group.

(ii) For all $v, v_1, v_2 \in V$ and for all $f, f_1, f_2 \in F$, we have

1. $f \bar{\cdot} (v_1 \bar{+} v_2) = f \bar{\cdot} v_1 \bar{+} f \bar{\cdot} v_2$.
2. $(f_1 + f_2) \bar{\cdot} v = f_1 \bar{\cdot} v \bar{+} f_2 \bar{\cdot} v$.
3. $(f_1 \cdot f_2) \bar{\cdot} v = f_1 \bar{\cdot} (f_2 \bar{\cdot} v)$.
4. $1 \bar{\cdot} v = v$.

**Example A.4** Let $(F, +, \cdot)$ be a field. Let $\mathbf{f} = (f_1, f_2, \ldots, f_n)$ and $\mathbf{g} = (g_1, g_2, \ldots, g_n)$ denote $n$-tuples where $f_i, g_j \in F$, $i, j \in (1, 2, \ldots, n)$. We say that $\mathbf{f}, \mathbf{g} \in F^n$, the set of $n$-tuples with components from $F$. Define component-wise addition

$$\mathbf{f} \bar{+} \mathbf{g} = (f_1 + g_1, f_2 + g_2, \ldots, f_n + g_n),$$

and component-wise scaler multiplication

$$\alpha \bar{\cdot} \mathbf{f} = (\alpha \cdot f_1, \alpha \cdot f_2, \ldots, \alpha \cdot f_n), \quad \alpha \in F.$$

Then set $F^n$ is a vector space over $(F, +, \cdot)$ under these operations. A field is a vector space over itself. The set of all infinite sequences with entries in a field $(F, +, \cdot)$ is a vector space under component-wise operations $(s_n) \bar{+} (t_n) = (s_n + t_n)$ and $f \bar{\cdot} (s_n) = (f \cdot s_n)$.

An expression of the form $f_1 \bar{\cdot} \mathbf{v}_1 \bar{+} f_2 \bar{\cdot} \mathbf{v}_2 \bar{+} \cdots \bar{+} f_r \bar{\cdot} \mathbf{v}_r$, where $f_i \in F$ and $\mathbf{v}_i \in V$ for all $i$, is called a *linear combination* of the vectors $\mathbf{v}_1, \mathbf{v}_2, \ldots, \mathbf{v}_r$. A set of $r$ vectors $\mathbf{v}_1, \mathbf{v}_2, \ldots, \mathbf{v}_r$. of $V$ are said to be *linearly independent* over $F$ if $\sum_{i=1}^{r} f_i \bar{\cdot} \mathbf{v}_i = 0$ implies that $f_i = 0$ for all $i \in (1, 2, \ldots, r)$. Otherwise, the set of vectors are linearly dependent.

Let $k$ be the largest integer for which $k$ independent vectors can be found in $V$. Then $k$ is the *dimension* of $V$ over $F$ and the set of $k$ independent vectors $\mathbf{v}_1, \mathbf{v}_2, \ldots, \mathbf{v}_k$ is called a *basis* of $V$ over $F$. Any vector $v \in V$ can be written as $v = \sum_{i=1}^{k} f_i \bar{\cdot} \mathbf{v}_i$ for $f_i \in F$ in a unique way.

## *Subspaces*

A subset of a vector space $V$ is a *subspace* if it is itself a vector space under the same operations defined for $V$. A set of vectors $\{\mathbf{v}_i\}$ *generates* a subspace $U$ of $V$, where any $\mathbf{u} \in U$ can be written as $\sum f_i \bar{\cdot} \mathbf{v}_i$, $f_i \in F$, for a finite number of nonzero $f_i$.

## *Orthogonal Complement*

The *inner product* of two vectors $\mathbf{u}, \mathbf{v} \in F^n$ is a scaler defined as follows:

$$\langle \mathbf{u}, \mathbf{v} \rangle = u_1 \cdot v_1 + \cdots + u_n \cdot v_n.$$

Two vectors are said to be *orthogonal* to each other if their inner product is zero. If a vector is orthogonal to all the vectors in a set, then it is said to be orthogonal to the set itself. In fact, if $U$ is a subspace of a vector space $V$, then the set of all vectors in $V$ that are orthogonal to $U$ is itself a subspace of $V$. The set of vectors orthogonal to $U$ is called the *orthogonal complement* (or the *dual space*) of $U$ and is denoted $U^{\perp}$.

# A.3 MATRICES

A *matrix* over a field $(F, +, \cdot)$ is a rectangular array of elements of the field. A $k \times n$ matrix $M$ has $k$ rows and $n$ columns, and is denoted

$$M = \begin{bmatrix} m_{11} & m_{12} & \cdots & m_{1,n} \\ m_{21} & m_{22} & \cdots & m_{2,n} \\ \vdots & \vdots & & \vdots \\ m_{k,1} & m_{k,2} & \cdots & m_{k,n} \end{bmatrix}.$$

The element in the $i$th row and $j$th column is denoted $m_{i,j}$ and is a member of the set $F$. A compact notation for a matrix is $M = [m_{i,j}]$.

Two $k \times n$ matrices $A = [a_{i,j}]$ and $B = [b_{i,j}]$ can be added as follows:

$$[a_{i,j}] + [b_{i,j}] = [a_{i,j} + b_{i,j}]$$

Matrix multiplication is a bit more involved. Two matrices can be multiplied only if the number of columns of the first matrix is equal to the number of rows of the second matrix. Let $A = [a_{i,j}]$ be a $k \times l$ matrix and let $B = [b_{i,j}]$ be a $l \times m$ matrix. Then, the product matrix $C = c_{i,j}$ is a $k \times m$ matrix, and each element $c_{i,j}$ is obtained as follows:

$$c_{i,j} = \sum_{r=0}^{r=m} a_{i,r} b_{r,j}$$

The set of all $k \times n$ matrices is an abelian group under matrix addition. The set of all $n \times n$ square matrices is a noncommutative ring under matrix addition and multiplication.

The $k$ rows of a $k \times n$ matrix can be viewed as $n$-tuples (or vectors) whose components are members of the field $(F, +, \cdot)$. The set of rows of a $k \times n$ matrix generates a vector space known as the *row space* of the matrix whose dimension is the number of linearly independent rows. Similarly, the $n$ columns of a $k \times n$ matrix can be viewed as $k$-tuples (or vectors) whose components are also members of the field $(F, +, \cdot)$. The vector space generated by the set of columns of a matrix is called the *column space* of the matrix, and its dimension is the number of linearly independent columns.

Let $R$ be a principal ideal domain. A matrix over $R$ is *unimodular* if it is invertible and its determinant is a unit in $R$.

**Definition A.1** *Two rectangular $k \times n$ matrices $A$ and $B$ over $R$ are said to be equivalent if there exist unimodular matrices $X$ and $Y$ such that $B = XAY$.*

Here, $X$ is a $k \times k$ matrix over $R$ and $Y$ is an $n \times n$ matrix over $R$, each with determinant equal to a unit in $R$.

The following are called *elementary* operations on a matrix: 1) interchanging two rows or two columns, 2) addition to a row of a multiple of another row, or addition to a column of a multiple of another column, and 3) multiplication of a row or a column by a nonzero constant. Two matrices are *equivalent* if one can be derived from the other by a finite number of elementary operations.

An elementary row operation can be carried out by performing that operation on the identity matrix and premultiplying by the result. Similarly, an elementary column operation can be carried out by performing that operation on the identity matrix and postmultiplying by the result. Since the matrices performing the operations are invertible, any of these elementary operation transforms a matrix into an equivalent one.

We now state the *invariant factor theorem* for matrices [96]. This theorem is used in Chapter 4 to obtain inverses of convolutional encoders.

**Theorem A.1 (Invariant Factor Theorem for Matrices)** *Let $G$ be a rectangular matrix with coefficients in $R$. Then, $G$ is equivalent to diag $\{\gamma_1, \gamma_2, \ldots, \gamma_r, 0, \ldots, 0\}$, where $\gamma_i \in R$, $\gamma_i \neq 0$, and $\gamma_i \mid \gamma_{i+1}$, $1 \leq i \leq r$. The elements $\{\gamma_i\}$ are called the invariant factors of $G$ and are determined up to unit factors.*

The invariant factors are unique and are given by $\gamma_i = \triangle_i/\triangle_{i-1}$, where $\triangle_i$ is the greatest common divisor of the $i \times i$ minors of $G$, and $\triangle_0 = 1$ by convention. If $G$ is a $k \times n$ matrix over $R$ with an invariant factor decomposition $A\Gamma B$, with $\Gamma$ a $k \times n$ diagonal matrix with diagonal elements $\gamma_i$, $1 \leq i \leq k$ called the invariant factors of $G$ with respect to $R$, then we have $r \leq k$, and $\gamma_i$ divides $\gamma_{i+1}$ if $\gamma_{i+1}$ is not zero.

The invariant factor theorem can be extended to rings of fractions as follows [147]: Let $R$ be a principal ideal domain and let $Q$ be a ring of fractions of $R$. Let $G$ be a $k \times n$ $Q$-matrix. Let $\psi$ be the least common multiple of all denominators in $G$; then $\psi G$ is an $R$-matrix. Consequently, $\psi G$ has an invariant factor decomposition $\psi G = A\Gamma' B$. Dividing through $\psi$, we obtain an invariant factor decomposition of the $Q$-matrix $G$ with respect to $R$ $G = A\Gamma B$ where $\Gamma = \Gamma'/\psi$. The matrices $A$ and $B$ are $R$-matrices and have $R$-matrix inverses. The diagonal elements $\gamma_i$ of $\Gamma$ are elements of $Q$ uniquely determined as $\gamma_i = \gamma_i'/\psi = \alpha_i/\beta_i$, where $\gcd(\alpha_i, \beta_i) = 1$. Since $\gamma_i' \mid \gamma_{i+1}'$ if $\gamma_{i+1}' \neq 0$, we have that $\alpha_i \mid \alpha_{i+1}$ if $\alpha_{i+1} \neq 0$ and $\beta_{i+1} \mid \beta_i$, $1 \leq i \leq k-1$.

Given a $k \times n$ matrix $G$ over $R$, the equivalent diagonal matrix $\Gamma$ is obtained by a sequence of elementary operations [185]. The sequence of row operations is cumulatively expressed as $A$, and the sequence of column operations is cumulatively expressed as $B$. The matrix $\Gamma$ is called the *Smith form* of $G$ [232].

# A.4 POLYNOMIALS

Let $(R, +, \cdot)$ be a ring. An expression of the following form is called a *polynomial*:

$$p(x) = \sum_{i=0}^{i=n} p_i \, x^i = p_0 + p_1 x + p_2 x^2 + \cdots + p_n x^n,$$

where $p_i \in R$, $i \in (0, 1, \ldots, n)$, and $x$ is not in $R$ and is called an *indeterminate* over $(R, +, \cdot)$. The coefficient $p_n$ of the highest power term of $x$ in $p(x)$ is called the *leading coefficient* of the polynomial, and the highest power $n$ is called the *degree* of $p(x)$. If all $p_i = 0$, then $p(x)$ is called the *zero* polynomial. The degree of the zero polynomial is defined to be $-\infty$. Polynomials in which $p_i = 0$ for $1 \leq i \leq n$ and $p_0 \neq 0$ are called *constant* polynomials. The degree of a constant polynomial is 0. If $p_n = 1$, then $p(x)$ is called a *monic* polynomial. Two polynomials $p(x)$ and $q(x)$ are said to be *equal* if $p_i = q_i$ for all $i$.

## A.4.1 Polynomial Rings

We can define addition of two polynomials $p(x) = \sum_{i=0}^{i=n} p_i \, x^i$ and $q(x) = \sum_{i=0}^{i=n} q_i \, x^i$ as the polynomial

$$p(x) + q(x) = \sum_{i=0}^{i=n} (p_i + q_i) \, x^i$$

and multiplication of two polynomials $p(x) = \sum_{i=0}^{i=m} p_i \, x^i$ and $q(x) = \sum_{i=0}^{i=n} q_i \, x^i$ as the polynomial

$$p(x) \cdot q(x) = \sum_{k=0}^{k=m+n} r_k x^k,$$

where

$$r_k = \sum_{\substack{i+j=k \\ 0 \leq i \leq m, 0 \leq j \leq m}} p_i \cdot q_j.$$

The set of all polynomials in $x$ over $(R, +, \cdot)$ is denoted $R[x]$. The set $R[x]$ forms a ring under polynomial addition and multiplication and is called the *polynomial ring* over $R$.

The set of all polynomials over a field $(F, +, \cdot)$ is denoted $F[x]$ and forms a ring under polynomial addition and multiplication. In this ring, polynomial division with remainder can be defined. Let $g(x) \neq 0$ be a polynomial in $F[x]$. Then any polynomial $f(x) \in F[x]$ can be written as

$$f(x) = q(x)g(x) + r(x),$$

where $q(x), r(x) \in F[x]$ and degree of $r(x)$ is less than the degree of $g(x)$. If $f(x) = q(x)g(x)$, then we say that $f(x)$ is a *multiple* of $g(x)$, or that $g(x)$ is a *divisor* of $f(x)$.

$F[x]$ is a principal ideal domain. It is very similar to the ring $(\mathcal{Z}, +, \cdot)$ of integers. A polynomial $p(x) \in F[x]$ with $p_i \neq 0$ for at least one $i > 0$ is called an *irreducible polynomial over* $F$ if whenever $p(x) = q(x)\ r(x)$, with $q(x),\ r(x) \in F[x]$, then either $q(x)$ is of degree 0 or $r(x)$ is of degree zero. Otherwise, it is called a *reducible* polynomial. Irreducible polynomials in $F[x]$ behave just like primes do in $(\mathcal{Z}, +, \cdot)$.

Let $p(x),\ q(x) \in F[x]$. A monic polynomial $d(x)$ is called a *greatest common divisor (gcd) of* $p(x)$ *and* $q(x)$ if $d(x)|p(x)$ and $d(x)|q(x)$, and whenever $c(x)|p(x)$ and $c(x)|q(x)$, for some $c(x) \in F[x]$, then $c(x)|d(x)$. The *gcd* of $p(x)$ and $q(x)$ is denoted as $(p(x),\ q(x))$. Two polynomials $p(x)$ and $q(x)$ are said to be *relatively prime* if $(p(x),\ q(x)) = 1$. In a principal ideal domain, irreducible and prime are the same.

For $f(x) \in F[x]$, the quotient ring $F[x]/(f(x))$ is a field if and only if $f(x)$ is irreducible over $F$. This is analogous to the quotient field $\mathcal{Z}/(n)$ where $n$ is a prime integer.

## A.4.2 Sequences

This section is based on a similar section in [362].

Consider a finite nonempty set $S$. A *sequence* over $S$ can be defined as a mapping from $\mathcal{Z}$ into $S$ given by $\mathcal{Z} \to S : i \to s_i$, and the set of all such sequences is denoted by $S^{\mathcal{Z}}$. An element of $S^{\mathcal{Z}}$ is of the form

$$s = (\ldots,\ s_{-i},\ s_{-i+1},\ \ldots,\ s_0,\ s_1,\ \ldots,\ s_i,\ \ldots).$$

An alternate representation for a sequence $s$ is a power series over $S$ in some indeterminate $D$, given by

$$\mathbf{s} = \sum_{i \in \mathcal{Z}} s_i D^i.$$

Let $S$ contain an element denoted by 0 and called zero. If all $s_i$ are zero, $s$ itself is said to be a zero sequence. For a nonzero $s \in S^{\mathcal{Z}}$, the smallest $i$ for which $s_i$ is nonzero is called the *delay* of $s$ and is denoted by $del(s)$. Similarly, for a nonzero $s \in S^{\mathcal{Z}}$, the largest $i$ for which $s_i$ is nonzero is called the *degree* of $s$ and is denoted by $deg(s)$. If in either case no such smallest (resp. largest) integer $i$ exists for which $s_i$ is nonzero, the delay (resp. degree) of $s$ is $-\infty$ (resp. $\infty$). In general, the $del(s)$ and $deg(s)$ can be positive, negative, or zero.

When $S$ contains a zero, the following subsets of $S^{\mathcal{Z}}$ are interesting. The set $S((D))$ of ***Laurent series*** over $S$ is the set of sequences $s$ in which only a finite number of coefficients $s_i$ with a negative index $i$ are nonzero. The set $S[D]$ of ***polynomials*** over $S$ is the set of sequences $s$ in which $s_i$ is zero for $i < 0$ and only a finite number of $s_i$ are nonzero.

Now we consider the case when $S$ can be assigned the structure of a field $(F, +, \cdot)$. We can define addition in $F((D))$ as

$$s + t = \sum_{i=r}^{\infty} (s_i + t_i)\, D^i,$$

where $s_i + t_i$ is the addition of $s_i$ and $t_i$ in $F$, and $r \leq min(del(s),\ del(t))$. The multiplication in $F((D))$ is defined as

$$s \cdot t = \sum_{i=r}^{\infty} \left(\sum_{j} s_{i-j} \cdot t_j\right) D^i,$$

where the summation in the parentheses is carried out in $F$. With these two operations, $F((D))$ has the structure of a field and is called the field of *formal Laurent series* over $F$. This is also the field of quotients of the ring $F[[D]]$ of *formal power series.* The set $F[D]$ of polynomials in $D$ over $F$ is a subset of $F((D))$. $F[D]$ has the structure of a ring. The field of quotients of the polynomials is the field $F(D)$ of *rational functions.* Note that $F(D) \subset F((D))$. An interesting subset of $F(D)$ is the set of ratios of polynomials in $D$ over $F$ with a nonzero constant term in the denominator. Elements in this set, denoted by $F_{rz}[D]$, are *realizable functions*, i.e., realizable by causal finite-state systems.

# B

# BLOCK CODES

Along with convolutional codes, block codes make up the two major types of channel codes. Block codes were invented by Hamming [181], and were used as examples in Shannon's original papers [413].[1] This appendix is a very brief introduction to binary linear block codes. We refer interested readers to many good and extensive textbooks that deal with the topic of block codes authoritatively and exhaustively [358, 39, 303, 285, 448, 447, 384, 8].

## B.1 DEFINITION, DUAL CODES, AND DISTANCE PROPERTIES

Let $F = GF(2)$ be the binary field, and let $F^n$ be the $n$-dimensional vector space of binary $n$-tuples.

**Definition B.1** *An $(n, k)$ binary linear block code $C$ is a $k$-dimensional linear subspace of $F^n$.*

The source information is divided into blocks of $k$ bits, which are then encoded independently into $n$ bits blocks. The information blocks are called *information words*. The coded blocks are called the *codewords* and $n$ is called the *block length* or *code length* or *length*. The ratio $k/n$ is called the *code rate*. A binary $k$-tuple $\mathbf{x}$ is encoded as the binary $n$-tuple $\mathbf{y} = \mathbf{x}\mathbf{G}$, where $\mathbf{G}$ is the generator matrix.

[1] Golay also proposed a block coding scheme in 1949 [164]. But Hamming's work on block coding was first, and was delayed in publication due to patent considerations [40].

**Definition B.2** *A generator matrix* $\mathbf{G}$ *for a block code* $C$ *is a* $k \times n$ *matrix the row space of which is* $C$.

If the generator matrix $\mathbf{G}_1$ of code $C_1$ is obtained from the generator matrix $\mathbf{G}_2$ of code $C_2$ by a sequence of elementary column operations of the following type: 1) transposition of two columns, and 2) multiplication of a column by a nonzero scalar, then $C_1$ and $C_2$ are equivalent. Any $(n,\ k)$ code is equivalent to one that has a generator matrix of the form $\mathbf{G} = [I_k \mid P]$, where $I_k$ is the $k \times k$ identity matrix and $P$ is a $k \times (n-k)$ matrix. $\mathbf{G}$ is in *standard form* (or *systematic form*) if $\mathbf{G} = [I_k \mid P]$.

**Definition B.3** *If* $C$ *is an* $(n,\ k)$ *block code, its dual code is*

$$C^{\perp} = \{\mathbf{y} \in F^n \mid \forall_{\mathbf{x} \in C}[< \mathbf{x}, \mathbf{y} >= 0]\}.$$

The generator matrix $\mathbf{H}$ of $C^{\perp}$ is also called the *parity check matrix* for $C$. The dual code $C^{\perp}$ is a linear $(n,\ n-k)$ code. If $\mathbf{G} = (I_k \mid P)$ is the generator matrix in standard form of the code $C$ then $\mathbf{H} = (-P^T \mid I_{n-k})$ is a generator matrix for $C^{\perp}$. In this case, we have $\mathbf{G}\mathbf{H}^T = 0$. If $C = C^{\perp}$ then $C$ is called a *self-dual* code. A vector $\mathbf{x} \in F^n$ is a codeword ($\mathbf{x} \in C$) if and only if $\mathbf{x}\mathbf{H}^T = \mathbf{0}$ (or if $\mathbf{H}\mathbf{x} = \mathbf{0}$).

**Definition B.4** *If* $C$ *is a linear code with parity check matrix* $\mathbf{H}$ *then for every* $\mathbf{x} \in \mathcal{A}^n$ *we call* $\mathbf{x}\mathbf{H}^T$ *the syndrome of* $\mathbf{x}$.

A nonzero syndrome sequence indicates presence of errors. Codewords are characterized by syndrome $\mathbf{0}$.

**Definition B.5** *The minimum distance (minimum weight) of a block code* $C$ *is*

$$d(C) = min\ \{d(\mathbf{x}, \mathbf{y}) | \mathbf{x} \in C,\ \mathbf{y} \in C,\ \mathbf{x} \neq \mathbf{y}\} = min\ \{w(\mathbf{x}) | \mathbf{x} \in C, \mathbf{x} \neq \mathbf{0}\}.$$

An $(n,\ k)$ linear code $C$ with minimum distance $d$ is called an $(n,\ k,\ d)$ code.

# B.2 DECODING

A maximum likelihood or nearest neighbor decoding scheme can be devised for block codes. Using this decoding scheme, it can be shown that an $(n,\ k,\ d)$ code $C$ corrects $t$ errors if $d \geq 2t+1$, or it detects $t$ errors if $d \geq t+1$.

## B.2.1 Coset Decoding

The algebraic structure of a block code can be put to use in devising an efficient nearest neighbor decoding scheme. Since $C$ is a subspace of $F^n$, it is also a subgroup of the additive group $(F^n,\ +)$. Then, for a given received word $\mathbf{r} \in F^n$, the set $\mathbf{r}+C$ is a coset of $C$. Now, if $\mathbf{e}$ is the error word corresponding to the received word $\mathbf{r}$, then using nearest neighbor decoding, it is most likely that $\mathbf{e}$ is an error word of minimum weight such that $\mathbf{r} - \mathbf{e} \in C$. The set $S_{\mathbf{r}} = \{\mathbf{e} : \mathbf{r} - \mathbf{e} \in C\}$ is set of all errors with respect to the received word $\mathbf{r}$ and hence contains the minimum weight error as well. Further reflection would reveal that the set $S_{\mathbf{r}}$ is the coset $\mathbf{r}+C$ of $C$.

The list of cosets of the code $C$ is called the *standard array* or the *decoding table* for $C$. Select as representative for each coset an element in that coset of minimum weight. Such a representative element is called the *coset leader.* The first row of the standard array is the code $C$. Each of the remaining rows is formed by selecting a minimum weight $n$-tuple not already in the preceding rows, and adding it to the entire first row (i.e., code $C$). The first column in the standard array consists of the coset leaders. Given a received word $\mathbf{r}$, a standard array is used along with the following decoding rule: *Locate* $\mathbf{r}$ *in the standard array and decode* $\mathbf{r}$ *to be the codeword at the top of the column in which* $\mathbf{r}$ *appears.* Note that this is the same as selecting the coset leader of the coset (row) in which $\mathbf{r}$ appears and subtracting it from $\mathbf{r}$. This decoding scheme is called *coset decoding.*

## B.2.2 Syndrome Decoding

Two $n$-tuples have the same syndrome if and only if they belong to the same coset of $C$. Since the rows of the standard array are cosets of $C$, a unique syndrome corresponds to each row. This suggests the following decoding procedure, known as *syndrome decoding*: 1) Compute the syndrome $\mathbf{rH}^T$ for a received word $\mathbf{r}$, 2) Determine the coset leader $\mathbf{e}$ that has the same syndrome as $\mathbf{r}$, 3) Decode $\mathbf{r}$ as $\mathbf{r} - \mathbf{e}$. This decoding scheme requires a *syndrome table* which stores syndromes and their corresponding coset leaders. A syndrome table is considerably smaller than the standard array, and does not require a search through the table.

### B.2.3 Probability of Error

Assuming a BSC with a transition probability $p$ and a $t$ error correcting block code used strictly for error correction, the probability that the decoder make an inaccurate decoding decision is upper bounded by

$$P(E) \leq \sum_{i=t+1}^{n} \binom{n}{i} p^i(1-p)^{n-i}. \tag{B.1}$$

The performance of a code depends upon the decoding scheme used at the receiver. If coset decoding is used, then an exact expression for the probability of decoding error can be obtained. For an $(n,\ k)$ linear code $C$, let $A_i$ be the number of codewords of weight $i$ in $C$. The numbers $A_0,\ A_1,\ \ldots,\ A_n$ are called the *weight enumerators* of $C$. Note that if the minimum distance of the code is $d$, then $A_1$ to $A_{d-1}$ are zero. Knowledge of these numbers allows us to compute the probability of decoding error. A decoding error occurs if and only if the error pattern is not a codeword. Hence, the error probability is

$$P(E) = 1 - \sum_{i=t+1}^{n} A_i\, p^i(1-p)^{n-i}. \tag{B.2}$$

If $C$ is used only for error detection, it is possible to determine the probability of undetected error $P_u(E)$. An undetected error occurs only if the errors on the channel change one codeword into another, i.e., if the error word itself is a codeword of $C$. Then

$$P_u(E) = \sum_{i=t+1}^{n} A_i\, p^i(1-p)^{n-i}. \tag{B.3}$$

It can be shown that the probability of undetected error is upper bounded by $2^{-(n-k)}$ [285], i.e.,

$$P_u(E) \leq 2^{-(n-k)}. \tag{B.4}$$

This shows that linear $(n,\ k)$ codes exist with $P_u(E)$ decreasing exponentially with the number of parity check digits $(n-k)$. This upper bound is quite small even for a moderate number of parity check digits.

# C

# COMBINED MODULATION AND CODING

The system resources available to a digital communication system designer are the signal power and the channel bandwidth. Given the channel noise characteristics, the design objective typically is to optimize the usage of these resources in maximizing the information throughput while striving to meet certain performance criterion such as the probability of error at a given signal-to-noise ratio. If more bandwidth is available and the signal power is limited, the desired probability of error can be achieved by using channel coding. On the other hand, if the bandwidth is limited, the signal power has to be increased to meet the performance requirements. This point of view assumes that channel coding and modulation are treated as separate components of the system and are separately optimized.

A recent and exciting application of convolutional coding has been in communication systems that combine channel coding and modulation in one step to achieve better performance with no bandwidth expansion. This approach was proposed by Ungerboeck [445, 442, 443, 444] and has been called *trellis-coded modulation* (TCM). Massey [313] pointed out that communication system designs that treat the coding and the modulation components of the system as being independent of each other do not achieve the best performance. Combined coding and modulation is based on performing channel coding directly in the modulation-signal space rather than in the baseband-signal space [426].

If channel coding and modulation are treated separately, then the source information stream is encoded using a channel encoder and and digital modulation is performed by using $\log_2 M$ encoded bits to select a signal waveform from a set of $M$ possible signals. Digital modulation involves mapping discrete symbols to a signal in a set or constellation. Common examples of digital modulation schemes are amplitude modulation with two or more signal points in the constellation (e.g., 2-AM, 4-AM), phase modulation with two or more signal points in the constellation of constant amplitude but different phases (e.g. 2-PSK (BPSK), 4-PSK (QPSK), 8-PSK, etc.), a combination of amplitude and phase modulation, i.e., multilevel/phase schemes, and other generalizations [266, 365]. For instance, in $M$-PSK, each set of $\log_2 M$ bits are used independently to select the signal to be sent over the waveform channel.

A given signal set is characterized by the minimum distance between two signal points. If the signal power is kept constant, then selecting a larger signal set (e.g., 8-PSK instead of 4-PSK) results in a reduction of the minimum distance. If signals are selected independent of each other, there is a net reduction in the performance of the scheme due to the reduced minimum distance and the probability of error increases. In TCM, coding redundancy is introduced directly in the selection of modulation signals, e.g., by using a convolutional encoder. Thus, using memory provides an increase in the minimum free distance. At the receiver, the convolutional decoder (e.g. Viterbi decoding) performs sequence decoding instead of symbol decoding. Quite simply, even though the signal points are closer to each other in the expanded signal set, the memory is used to make consecutive selection of signal points very close to each other less likely, thereby effectively increasing the distance between sequences of signal points.

Since Ungerboeck's work, a large body of research has been devoted to TCM, primarily due to its applicability in efficient modem designs. Basic ideas of TCM are explained in tutorial articles by Ungerboeck [443, 444], Sklar [426], and Massey [316]. Use of TCM in modem designs is described by Forney *et al.* [157]. TCM has been applied to many other channels besides the telephone voice channel such as the partial-response channels (used in magnetic recording) and the fading channels (used in mobile communications) [51, 266, 425, 365, 17].

# D

# CONCATENATED CODING

An attempt to achieve very low error probabilities using a single code requires very long codes, and consequently, very complex coding systems. One way around this problem is to use two or more codes that are not very long in a manner that reduces the complexity considerably without increasing the error rate significantly. Elias [128] proposed an *iterated* coding scheme based on repeated use of simple codes. Forney's monograph [144] describes a technique known as *concatenated coding* that uses two or more block codes of lesser complexity in a cascaded manner to achieve very high reliability. The general model of a concatenated coding system is shown in Figure D.1.

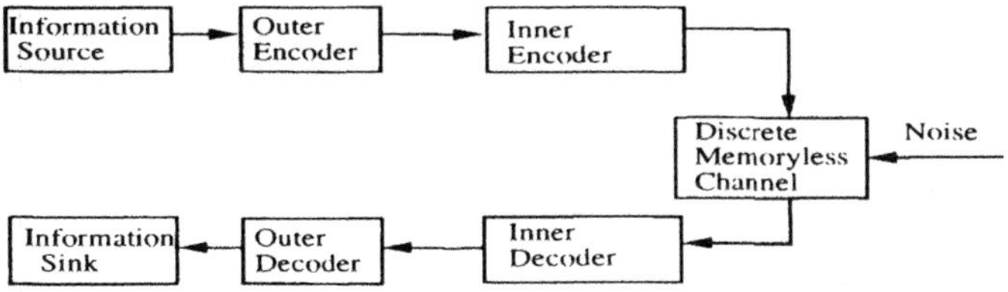


**Figure D.1** A concatenated coding system.

Forney briefly mentioned the possible use of a convolutional encoder as the inner encoder with sequential decoder as the inner decoder. This idea was further investigated by Falconer [131, 132] and Jelinek and Cocke [219]. The use of a Viterbi decoder as the inner decoder was suggested by Odenwalder [345]. Odenwalder also suggested the use of interleaving between the inner and outer encoders to effectively randomize the decoded error bursts introduced by the inner decoder. The resulting concatenated system is shown in Figure D.2. Zeoli [491] suggested the use of reliability information generated by the

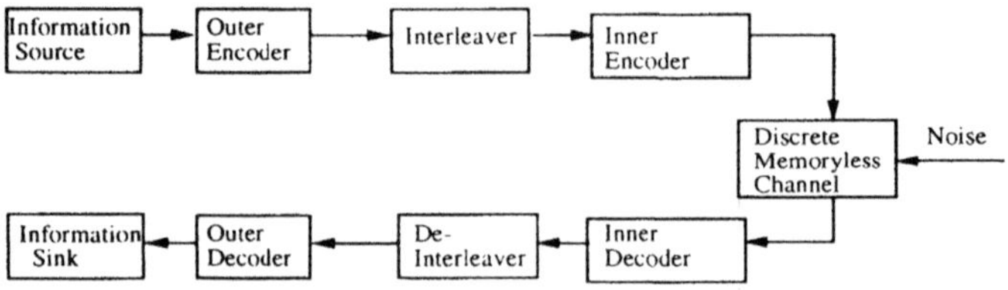


**Figure D.2** A concatenated coding system with interleaving.

inner decoder to improve the overall performance. The inner decoder provides reliability information about the decoded bits in addition to the decoded bits. This *side information* is used by the outer decoder to make a better decoding decision, thereby improving the performance of the scheme. A similar scheme was also proposed by Jelinek [218]. Lee [268, 270] suggested another technique, based on byte-oriented unit-memory convolutional codes, for extracting reliability information from the inner decoder. The capacity and cutoff rate of a concatenated coding system, with or without interleaving, and with or without side information has been evaluated by Herro, Costello, and Hu [201].

One of the main applications of concatenated coding has been in hybrid ARQ systems. Recall the VR-HARQ schemes of Chapter 10. The high-rate block code is first used to encode a $k$-bit message into an $n$-bit codeword. This $n$-bit codeword is then encoded using a rate one-half convolutional encoder to obtain two $(n+m)$-bit blocks, called the information block and parity block, where $m$ is the memory order of the convolutional encoder. We can treat the block code as the outer code and the convolutional code as the inner code. When at least two transmissions are required, the inner convolutional decoder is used to correct the errors that may have occurred in the two transmissions, and then the outer block decoder is used to perform error detection. If no errors are detected, then the $k$-bit estimate produced by the outer decoder is delivered to the user. It is clear from this description that this scheme is very much like a concatenated coding scheme. For a variety of results on the performance of concatenated hybrid ARQ schemes the readers may referred to [247, 104, 283, 52, 103].

Other work related to concatenated coding is described in [335], [478, 292, 179, 277, 408, 254, 283, 405, 52, 103]. Practical applications of concatenated coding are summarized in [41, 42] and in [324, 365].

# E

# NOTES ON THE REFERENCES

This appendix is a guide to over 450 research articles, books, technical reports, dissertations, etc. listed as references in this book. Research articles on convolutional codes have mostly appeared in the following journals: IEEE Transactions on Information Theory, IEEE Transactions on Communications, IEEE Journal on Selected Areas in Communications, and IEEE Communications Magazine. Many international conferences that devote sessions to research on convolutional codes are as follows: IEEE International Symposium on Information Theory (ISIT), IEEE International Communications Conference (ICC), IEEE Military Communications Conference (MILCOM), IEEE INFOCOM, IEEE GLOBECOM, IEEE Vehicular Technology Conference (VTC), and International Zurich Seminar on Digital Communications. Many other journals and conferences, besides the ones listed here, have included research dealing with convolutional codes.

## *Books*

Besides the present book, Piret's book [362] is the only one devoted entirely to convolutional codes. It deals mainly with the algebraic aspects of convolutional codes. Many coding theory books have devoted more than one chapters to convolutional codes; among them are the ones by Peterson [358], Massey [308], Lin [284], McEliece [324], Viterbi and Omura [454], Bhargava *et al.* [48], Clark and Cain [82], Blahut [56], Lin and Costello [285], Michelson and Levesque [330], Wiggert [470], Rhee [373], Anderson and Mohan [17], Sweeney [436], and Wicker [469].

Single chapter (or a part of a chapter) devoted to convolutional codes appear in the books by Fano [134], Wozencraft [474], Wozencraft and Jacobs [473],

Berlekamp [39], Gallager [159], Kuo [258], Massey [317], Arazi [19], Haykin [191], Lee and Messerschmitt [266], Sklar [426], Proakis [365], Imai [213], Nussbaumer [343], Adamek [8], Biglieri *et al.* [51], Hoffman *et al.* [204], and van Lint [447]. Collection of papers that include papers on convolutional codes have been edited by Berlekamp [40], Longo [294], and Viterbi [455].

## *Dissertations*

A partial list of doctoral dissertations dealing with convolutional codes is as follows: (1950's) [475]; (1960's) [431, 432, 131, 91]; (1970's) [160, 345, 386, 420, 127, 78, 268, 198, 360, 182]; (1980's) [119, 86, 296, 332, 69, 300, 83]; (1990's) [293, 109, 306, 353].

## *Tutorial and Survey Articles*

Convolutional codes were first proposed in [129]. A survey of coding theory appeared in [471]. Applications of convolutional codes have been discussed in [146, 216, 41, 42, 479]. surveyed alternatives for digital communications. Tutorials on digital communications that discuss convolutional codes are [376, 46, 423, 424]. Convolutional decoding techniques are reviewed in [170].

## *Part II: Definitions, Structure, and Classifications*

- Definitions and Algebraic Structure

  [129] [480] [321, 322, 392] [349] [147, 149, 153] [24] [387] [337] [490] [117] [102] [421] [315, 317] [154] [225] [326] [158]

- Distance Properties

  Column distance function, Free distance, Minimum distance: [311] [92, 91] [331] [22, 260] [212] [495, 2, 363, 47, 77], Distance profile, Distance Spectrum, Decision Depth: [221, 222, 223, 224] [89] [174] [70], Distance bounds: [378, 195, 377, 456, 194, 93, 1].

- Classes of Convolutional Codes and Encoders

  - Fixed and time-varying: [315, 356], systematic and nonsystematic: [92, 318], degenerate and nondegenerate: [24], catastrophic and non-catastrophic: [322].
  - Minimal: [147, 149, 153, 421, 225]
  - Locally invertible: [54, 115, 111, 110, 53]; [112, 114, 113, 116].

- Complementary, QLI, ELI: [23, 24] [318, 192, 193]
- Punctured: [379] [68] [205, 206, 208, 207] [487, 486] [273] [26, 171, 27] [177, 178, 180] [329] [124]; [176, 341, 241, 238].
- Unit-memory: [269] [263] [439, 230] [3, 5, 4] [85, 108, 228, 391]
- Uniform: [309, 379], self-orthogonal: [476, 477, 478].
- Tail biting: [301, 302, 429, 458, 457]
- Miscellaneous: [38, 433, 126] [229, 359, 102, 130, 355, 385] [200, 382, 406, 361, 118, 105]

■ Listings of Convolutional Encoders (see Section 4.12): [64, 287] [261] [269] [79] [175] [221, 222, 223, 224] [354] [192] [100] [271, 272] [15, 13] [70] [388] [139] [435] [489]

■ Convolutional Code Construction: [319] [226, 227] [458] [488, 489]

## *Part III: Decoding Techniques*

■ Viterbi Decoding

- Algorithm and performance: [451] [453, 350] [452, 150, 151] [197] [446, 364] [162] [199, 351]
- VLSI Architectures: [140, 142, 141, 143] [84] [167, 168, 166] [59] [73] [255, 214, 107] [418, 419] [334] [295] [390] [372] [282] [438] [65]
- Variations on Viterbi Decoding: [169] [441, 368] [281] [179] [485] [407] [97] [462] [404]

■ Sequential Decoding

[474, 135, 494, 217] [172] [80] [14] [357] [148] [156] [265] [318] [299] [449] [396] [215] [161] [312, 496] [152] [81] [16] [278] [336] [275]

■ Majority-Logic Decoding

[308] [389] [314] [397] [437, 173, 264]

■ Table Look-Up Decoding

[196, 454, 82, 56] [54] [115, 111, 110] [251, 250] [112, 114, 113] (Feedback decoding) [196, 308] (Definite decoding) [377]

■ Other Decoding Schemes: [28] [267, 401, 402, 403] [94] [340, 187] [369, 370] [371] [220] [280] [210] [428] [256] [352]

- Probability of Error: [414, 415] [145] [58] [61] [183] [66] [400] [259] [472] [71] [202] [417] [62]
- Error Propagation: [320, 380, 377] [310] [434] [339] [67] [274]

## *Part IV: Applications*

- Fixed-Redundancy Error Control Schemes
  - Error Control Tutorials: [35, 36, 63, 286]
  - Conventional ARQ [399, 398] [393] [307] [136] [60] [440] [10] [72] [121] [18] [289] [11] [461] [333] [33] [34]
  - Variations on Pure ARQ Schemes: [29, 30, 31] [6, 7] [243]
  - Type-I Hybrid-ARQ: [381] [394] [90] [481] [120, 122] [276] [409, 410] [239] [138] [137] [463, 466] [188, 189] [98, 99] [482, 483] [374, 375] [493] [395, 257, 103, 32]
  - ARQ with memory and code combining: [422] [74, 75, 76] [47]
  - Multiple copy: [461]
- Variable-Redundancy Error Control Schemes
  - Basic VR-HARQ: [101, 327].
  - Class A VR-HARQ: [290, 291, 288, 459, 460, 297, 298, 87, 88, 468, 244, 116].
  - Class B VR-HARQ: [304, 12, 176, 241, 235, 237].
  - Using block codes: [101, 304, 327, 291, 460, 328, 253, 125, 338, 252, 283, 468, 346].
  - Using convolutional codes: [12, 327, 288, 297, 460, 76, 176, 298, 241, 234, 341, 235, 237, 116].
  - Miscellaneous VR-HARQ: [305] [240] [242] [234] [236] [238] [233] [464] [467] [106] [25] [190] [484]
- Error Recovery in High-Speed Networks

  [44] [21, 20, 248, 323, 348, 347] [411, 412, 492] [262] [50] [49]

# REFERENCES


[1] M.J. Aaltonen. Linear programming bounds for tree codes. *IEEE Trans. Inf. Theory*, IT-25(1):85–90, 1979.

[2] K.A.S. Abdel-Ghaffar. Some convolutional codes whose free distances are maximal. *IEEE Trans. Inf. Theory*, IT-35(1):188–191, 1989.

[3] K.A.S. Abdel-Ghaffar. The number of convolutional codes of span one. *Comp. Math. with Applns.*, 21:33–38, 1991.

[4] K.A.S. Abdel-Ghaffar. On unit constraint-length convolutional codes. *IEEE Trans. Inf. Theory*, IT-38(1):200–206, 1992.

[5] K.A.S. Abdel-Ghaffar, R.J. McEliece, and G. Solomon. Some partial-unit-memory convolutional codes. In *Proc. IEEE Intl. Symp. Info. Th.*, page 196, 1991.

[6] F. Adachi and S. Ito. Efficient ARQ with time diversity reception technique - a time diversity ARQ. *Elec. Let.*, 22:1257–1258, 1986.

[7] F. Adachi, S. Ito, and K. Ohno. Performance analysis of a time diversity ARQ in land mobile radio. *IEEE Trans. Comm.*, 37(2):177–183, 1989.

[8] J. Adamek. *Foundations of Coding: Theory and Application of Error-Correcting Codes with an Introduction to Cryptography and Information Theory.* John Wiley and Sons, New York, 1991.

[9] A.O. Allen. *Probability, Statistics, and Queueing Theory.* Academic Press, Orlando, FL, 1978.

[10] M.H. Ammar and L.-R. Wu. Improving the output of point-to-multipoint ARQ protocols through destination set splitting. In *Proc. IEEE INFOCOM Conf.*, pages 262–271, 1992.

[11] M.E. Anagnostou and E.N. Protonotarios. Performance analysis of the selective repeat ARQ protocol. *IEEE Trans. Comm.*, COM-34(2):127–135, 1986.

[12] T.C. Ancheta. Convolutional parity check automatic repeat request. In *Book Abstr. IEEE Intl. Symp. Info. Th.*, page 150, 1979.

[13] J.B. Anderson. Limited search trellis decoding of convolutional codes. *IEEE Trans. Inf. Theory*, IT-35(5):944–955, 1989.

[14] J.B. Anderson. Sequential coding based on an error criterion. *IEEE Trans. Inf. Theory*, 38(3):987–1001, 1992.

[15] J.B. Anderson and K. Balachandran. Decision depths of convolutional codes. *IEEE Trans. Inf. Theory*, IT-35(2):455–459, 1989.

[16] J.B. Anderson and S. Mohan. Sequential coding algorithms: A survey and cost analysis. *IEEE Trans. Comm.*, COM-32(2):169–176, 1984.

[17] J.B. Anderson and S. Mohan. *Source and Channel Coding: An Algorithmic Approach.* Kluwer Academic Publishers, Boston, 1991.

[18] B. Arazi. Improving the throuhgput of an ARQ stop and wait scheme for burst noise channels. *IEEE Trans. Comm.*, pages 661–663, 1976.

[19] B. Arazi. *A Commonsense Approach to Error Correcting Codes.* M.I.T. Press, Cambridge, MA, 1988.

[20] E. Ayanoğlu and R.D. Gitlin. Performance improvement in broadband networks using forward error correction. In *Proc. Conf. Info. Sci. Sys.*, Princeton, NJ, 1992.

[21] E. Ayanoğlu, R.D. Gitlin, P.K. Johri, and W.S. Lai. Protocols for loss recovery in high-speed networks. In *Proceedings of the 7th International Teletraffic Congress Seminar*, Morristown, NJ, 1990.

[22] L.R. Bahl, C.D. Cullum, W.D. Frazer, and F. Jelinek. An efficient algorithm for computing free distance. *IEEE Trans. Inf. Theory*, IT-18(3):437–439, 1972.

[23] L.R. Bahl and F. Jelinek. Rate 1/2 convolutional codes with complementary generators. *IEEE Trans. Inf. Theory*, IT-17(6):718–727, 1971.

[24] L.R. Bahl and F. Jelinek. On the structure of rate 1/n convolutional codes. *IEEE Trans. Inf. Theory*, IT-18(1):192–196, 1972.

[25] S. Bakhtiyari, S. Kallel, and V. Leung. A robust type II hybrid ARQ scheme with code combining for mobile communications. In *IEEE Pac. Rim Conf. C.C.S.P.*, pages 214–217, 1993.

[26] G. Begin and D. Haccoun. High-rate punctured convolutional codes: Structure properties and construction technique. *IEEE Trans. Comm.*, COM-37(12):1381–1385, 1989.

[27] G. Begin, D. Haccoun, and C. Paquin. Further results on high-rate punctured convolutional codes for Viterbi and sequential decoding. *IEEE Trans. Comm.*, COM-38(11):1922–1928, 1990.

[28] J. Belzile and D. Haccoun. Bidirectional breadth-first algorithms for the decoding of convolutional codes. *IEEE Trans. Comm.*, 41(2):370–380, 1993.

[29] G. Benelli. An ARQ scheme with memory and soft error detectors. *IEEE Trans. Comm.*, COM-33(3):285–288, 1985.

[30] G. Benelli. An ARQ scheme with memory and integrated modulation. *IEEE Trans. Comm.*, COM-35(7):689–697, 1987.

[31] G. Benelli. A new method for the integration of modulation and channel coding in an ARQ protocol. *IEEE Trans. Comm.*, COM-40(10):1593–1606, 1992.

[32] G. Benelli. New ARQ protocols using concatenated codes. *IEEE Trans. Comm.*, 41(7):1013–1019, 1993.

[33] G. Benelli. A selective ARQ protocol with a finite-length buffer. *IEEE Trans. Comm.*, 41(7):1102–1111, 1993.

[34] G. Benelli. Some ARQ protocols with finite receiver buffer. *IEEE Trans. Comm.*, 41(4):513–523, 1993.

[35] R.J. Benice and A.H. Frey, Jr. An analysis of retransmission systems. *IEEE Trans. Comm.*, COM-12:135–145, 1964.

[36] R.J. Benice and A.H. Frey, Jr. Comparison of error control techniques. *IEEE Trans. Comm.*, COM-12:146–154, 1964.

[37] T. Berger. *Rate Distortion Theory.* Prentice-Hall, Englewood Cliffs, NJ, 1971.

[38] E.R. Berlekamp. A class of convolutional codes. *Inf. and Ctrl.*, 6:1–13, 1963.

[39] E.R. Berlekamp. *Algebraic Coding Theory.* McGraw-Hill, New York, 1968.

[40] E.R. Berlekamp, editor. *Key Papers in the Development of Coding Theory.* IEEE Press, New York, 1974.

[41] E.R. Berlekamp. The technology of error-correctng codes. *Proc. IEEE*, 68(5):564–593, 1980.

[42] E.R. Berlekamp, R.E. Peile, and S.P. Pope. The applications of error control to communications. *IEEE Comm. Mag.*, 25(4):44–57, 1987.

[43] D. Bertsekas and R. Gallager. *Data Networks.* Prentice-Hall, Englewood Cliffs, NJ, 1992.

[44] A. Bhargava and M.G. Hluchyj. Frame losses due to buffer overlows in fast packet networks. In *Proc. IEEE INFOCOM Conf.*, pages 132–139, 1990.

[45] A. Bhargava, J.F. Kurose, D. Towsley, and G. Vanleemput. Comparison of error control schemes in high-speed computer communication networks. *IEEE J. Sel. Areas in Comm.*, 6(9):1565–1575, 1988.

[46] V.K. Bhargava. Forward error correction schemes for digital communications. *IEEE Comm. Mag.*, 21(1):11–19, 1983.

[47] V.K. Bhargava, I.F. Blake, and L. Mason. Coding, diversity, and minimum distance. *Can. J. Elec. and Comp. Eng.*, 16(3):87–92, 1991.

[48] V.K. Bhargava, D. Haccoun, R. Matyas, and P.P. Nuspl. *Digital Communications by Sattellite.* John Wiley and Sons, New York, 1981.

[49] E.W. Biersack. Performance evaluation of forward error correction in atm networks. In *Proc. ACM SIGCOMM*, pages 248–257, Baltimore, MD, 1992.

[50] E.W. Biersack. A simulation study of forward error correction in ATM networks. *Computer Communications Review*, 22(1):36–47, 1992.

[51] E. Biglieri, D. Divaslar, P.J. McLane, and M.K. Simon. *Introduction to Trellis-Coded Modulation with Applications.* Macmillan, New York, 1991.

[52] A. Bigloo, Q. Wang, A. Gulliver, and V.K. Bhargava. A rate-adaptive concatenated coding scheme. In *IEEE Pac. Rim Conf. C.C.S.P.*, pages 85–88, 1993.

[53] D.L. Bitzer, A. Dholakia, H. Koorapaty, and M.A. Vouk. Rate-1/n locally invertible convolutional encoders. To be presented at the IEEE Intl. Symp. Info. Th., 1994.

[54] D.L. Bitzer and M.A. Vouk. A table-driven (feedback) decoder. In *Proc. Intl. Phoenix Conf. on Comp. Comm.*, pages 385–392, Phoenix, AZ, 1991.

[55] D.L. Bitzer, M.A. Vouk, A. Dholakia, E. Gonzalez, L.F. Wang, V. Srinivasan, T.M. Lee, S. Lo, and H. Koorapaty. Improvements to system and method for decoding of convolutionally encoded data. Application for U.S. Patent, 1992.

[56] R.E. Blahut. *Theory and Practice of Error Control Codes.* Addison-Wesley, Reading, MA, 1983.

[57] J.-Y. Le Boudec. The asynchronous transfer mode: A tutorial. *Comp. Net. ISDN Sys.*, 24:279–309, 1992.

[58] E.K. Bower and III S.J. Dwyer. A strengthened asymptotic bound for convolutional codes. *IEEE Trans. Inf. Theory*, IT-:433–435, 1969.

[59] M.A. Bree, D.E. Dodds, R.J. Bolton, S. Kumar, and B L.F. Daku. A modular bit-serial architecture for large-constraint-length Viterbi decoding. *IEEE Journal of Solid-State Circuits*, 27(2):184–189, 1992.

[60] H. Bruneel and M. Moeneclaey. On the throughput performance of some continuous ARQ strategies with repeated transmissions. *IEEE Trans. Comm.*, COM-34(3):244–249, 1986.

[61] E.A. Bucher and J.A. Heller. Error probability bounds for systematic convolutional codes. *IEEE Trans. Inf. Theory*, IT-16(2):219–224, 1970.

[62] A.G. Burr. Bounds and approximations for the bit error probability of convolutional codes. *Elec. Let.*, 29(14):1287–1288, 1993.

[63] H.O. Burton and D.D. Sullivan. Errors and error control. *Proc. IEEE*, 60(11):1293–1301, 1972.

[64] J.J. Bussgang. Some properties of binary convolutional code generators. *IEEE Trans. Inf. Theory*, IT-11(1):90–100, 1965.

[65] H.A. Bustamante, I. Kang, C. Nguyen, and R.E. Peile. Stanford telecom VLSI design of a convolutional decoder. In *Proc. IEEE MILCOM Conf.*, pages 171–178, 1989.

[66] J.B. Cain and R.W. Boyd. Convolutional code performance with psk signaling in nonstationary gaussian noise. In *National Teletraffic Conference*, pages 2.5.1–2.5.6, 1978.

[67] J.B. Cain and G.C. Clark. Some results on the error propagation of convolutional feedback decoders. *IEEE Trans. Inf. Theory*, IT-:681–683, 1972.

[68] J.B. Cain, G.C. Clark, Jr., and J.M. Geist. Punctured codes of rate (n-1)/n and simplified maximum likelihood decoding. *IEEE Trans. Inf. Theory*, IT-25:97–100, 1979.

[69] M. Cedervall. *Contributions to the Decoding and Structure of Convolutional Codes.* PhD thesis, Univ. of Lund, Lund, Sweden, 1983.

[70] M. Cedervall and R. Johannesson. A fast algorithm for computing distance spectrum of convolutional codes. *IEEE Trans. Inf. Theory*, IT-35(6):1146–1159, 1989.

[71] M. Cedervall, R. Johannesson, and K. Sh. Zigangirov. A new upper bound on the first-event error probability for maximum-likelihood decoding of fixed binary convolutional codes. *IEEE Trans. Inf. Theory*, IT-30(5):762–766, 1984.

[72] S. Ram Chandran and S. Lin. Selective-repeat-ARQ schemes for broadcast links. *IEEE Trans. Comm.*, COM-40(1):12–19, 1992.

[73] C.-Y. Chang and K. Yao. Systolic array processing of the Viterbi algorithm. *IEEE Trans. Inf. Theory*, 35(1):76–86, 1989.

[74] D. Chase. Code combining. In *Proc. IEEE MILCOM Conf.*, pages 71–77, 1984.

[75] D. Chase. Code combining- a maximum- likelihood decoding approach for combining an arbitrary number of noisy packets. *IEEE Trans. Comm.*, COM-85(5):385–393, 1985.

[76] D. Chase, P.D. Muellers, and J.K. Wolf. Application of code combining to a selective- repeat ARQ link. In *Proc. IEEE MILCOM Conf.*, pages 247–252, 1985.

[77] V.V. Chepyzhov, Ben J.M. Smeets, and K.S. Zigangirov. The free distance of fixed convolutional rate 2/4 codes meets the costello bound. *IEEE Trans. Inf. Theory*, IT-38(4):1360–1366, 1992.

[78] P.R. Chevillat. *Fast Sequential Decoding and a New Complete Decoding Algorithm.* PhD thesis, I.I.T., Chicago, IL, 1976.

[79] P.R. Chevillat and D.J. Costello, Jr. Distance and computation in sequential decoding. *IEEE Trans. Comm.*, COM-24(4):440–447, 1976.

[80] P.R. Chevillat and D.J. Costello, Jr. A multiple stack algorithm for erasurefree decoding of convolutional codes. *IEEE Trans. Comm.*, COM-25(12):1460–1470, 1977.

[81] P.R. Chevillat and D.J. Costello, Jr. An analysis of sequential decoding for specific time-invariant convolutional codes. *IEEE Trans. Inf. Theory*, IT-24(4):443–451, 1978.

[82] G.C. Clark, Jr. and J.B. Cain. *Error-Correction Coding for Digital Communications.* Plenum Press, New York, 1981.

[83] O. M. Collins. *Coding Beyond the computational Cutoff Rate.* PhD thesis, California Inst. Technol., California, 1989.

[84] O.M. Collins. The subtleties and intricacies of building a constraint length 15 convolutional decoder. *IEEE Trans. Comm.*, 40(12):1810–1819, 1992.

[85] O.M. Collins and M. Hizlan. Determinate state convolutional codes. *IEEE Trans. Comm.*, 41(12):1785–1794, 1993.

[86] R.A. Comroe. *An Analysis of ARQ Schemes for Land Mobile Radio Systems.* PhD thesis, I.I.T., Chicago, IL, 1981.

[87] R.A. Comroe and D.J. Costello, Jr. An analysis of ARQ schemes for data transmission in mobile radio systems. In *Proc. IEEE Intl. Comm. Conf.*, pages 5B.5.1–5B.5.5, 1982.

[88] R.A. Comroe and D.J. Costello, Jr. ARQ schemes for data transmission in mobile radio systems. *IEEE J. Sel. Areas in Comm.*, SAC-2(4):472–481, 1984.

[89] J. Conan. The weight spectra of some short low-rate convolutional codes. *IEEE Trans. Comm.*, COM-32(9):1050–1053, 1984.

[90] J. Conan and D. Haccoun. High speed transmission of reliable data on satellite channels. In *Proc. Fourth Intl. Conf. Dig. Sat. Comm.*, pages 269–274, 1978.

[91] D.J. Costello, Jr. *Construction of Convolutional Codes for Sequential Decoding.* PhD thesis, University of Notre Dame, Notre Dame, IN, 1969.

[92] D.J. Costello, Jr. A construction technique for random-error-correcting convolutional codes. *IEEE Trans. Inf. Theory*, IT-15(5):631–636, 1969.

[93] D.J. Costello, Jr. Free distance bounds for convolutional codes. *IEEE Trans. Inf. Theory*, IT-20(3):356–365, 1974.

[94] D.J. Costello, Jr. and T.N. Morrissey. Strengthened lower bound on definite decoding minimum distance for periodic convolutional codes. *IEEE Trans. Inf. Theory*, IT-:212–214, 1971.

[95] C.W. Curtis. *Linear Algebra: An Introductory Approach.* Allyn and Bacon, Boston, 1974.

[96] C.W. Curtis and I. Riener. *Representational Theory of Finite Groups and Associative Algebras.* John Wiley and Sons, New York, 1962.

[97] R. Cypher and C. B. Shung. Generalized trace back techniques for survivor memory management in the Viterbi algorithms. In *Proc. IEEE GLOBECOM Conf.*, pages 1318–1322, 1990.

[98] M.M. Darmon and P.R. Sadot. A hybrid FEC-ARQ communication system for zero-error digital transmission on the troposcatter channel. In *Proc. IEEE MILCOM Conf.*, pages 93–97, 1990.

[99] M.M. Darmon and P.R. Sadot. A hybrid FEC-ARQ communication system using sequential decoding. In G. Cohen and P. Charpin, editors, *Lec. Notes in Comp. Sci.*, volume 514, pages 378–384. Springer-Verlag, 1990. EUROCODE '90 Proceedings.

[100] D.G. Daut, J.W. Modestino, and L.D. Wismer. New short constraint length convolutional code constructions for selected rational rates. *IEEE Trans. Inf. Theory*, IT-28(5):794–800, 1982.

[101] G.I. Davida and S.M. Reddy. Forward error correction with decision feedback. *Inf. and Ctrl.*, 21:117–133, 1972.

[102] P. Delsarte and Ph. Piret. Semiregular convolutional codes: Definition, structure, and examples. *Inf. and Ctrl.*, 33:56–71, 1977.

[103] R.H. Deng. Hybrid ARQ schemes for point-to-multipoint communication over nonstationary broadcast channels. *IEEE Trans. Comm.*, 41(9):1379–1387, 1993.

[104] R.H. Deng and D.J. Costello, Jr. Reliability and throughput analysis of a concatenated coding scheme. *IEEE Trans. Comm.*, COM-35(7):698–705, 1987.

[105] R.H. Deng, Y.X. Li, and M.A. Herro. DC-free error-correcting convolutional codes. *Elec. Let.*, 29(22):1910–1911, 1993.

[106] R.H. Deng and H. Zhou. An adaptive coding scheme for mobile radio systems. In *Proc. IEEE Veh. Tech. Conf.*, pages 906–909, 1992.

[107] S. Deng, S. Kato, and T. Ishitani. Novel Viterbi decoder VLSI implementation and its performance. *IEEE Trans. Comm.*, 41(8):1170–1178, 1993.

[108] U. Dettmar and U.K. Sorger. New optimal partial unit memory codes based on extended BCH codes. *Elec. Let.*, 29(23):2024–2025, 1993.

[109] A. Dholakia. *Locally Invertible Convolutional Encoders, Table-Based Decoding, and Their Applications to High Speed Communications.* PhD thesis, ECE, NCSU, 1993.

[110] A. Dholakia, D.L. Bitzer, and M.A. Vouk. On a class of one-half rate noncatastrophic convolutional codes. Submitted to IEEE Trans. on Info. Th.

[111] A. Dholakia, M.A. Vouk, and D.L. Bitzer. Table-based decoding of rate one-half convolutional codes. Submitted to IEEE Trans. on Comm.

[112] A. Dholakia, M.A. Vouk, and D.L. Bitzer. High speed table-driven correction and decoding in convolutionally encoded type-I hybrid-ARQ protocols. In *Proc. IEEE MILCOM Conf.*, Boston, MA, 1993.

[113] A. Dholakia, M.A. Vouk, and D.L. Bitzer. A lost cell recovery technique using convolutional coding at the ATM adaptation layer in B-ISDN/ATM. In *Proc. Fifth Intl. Conf. Data Comm. Sys. Perf.*, 1993.

[114] A. Dholakia, M.A. Vouk, and D.L. Bitzer. A lost packet recovery technique using convolutional coding in high speed networks. In *Proc. IEEE MILCOM Conf.*, Boston, MA, 1993.

[115] A. Dholakia, M.A. Vouk, and D.L. Bitzer. Table-driven decoding of binary one-half rate nonsystematic convolutional codes. In *Proc. IEEE Intl. Symp. Info. Th.*, page 270, 1993.

[116] A. Dholakia, M.A. Vouk, and D.L. Bitzer. A variable redundancy hybrid ARQ scheme using invertible convolutional codes. To be presented at the IEEE Veh. Tech. Conf., 1994.

[117] B.W. Dickinson. A new characterization of canonical convolutional encoders. *IEEE Trans. Inf. Theory*, IT-22(3):352–354, 1976.

[118] S. Dolinar, R. McEliece, F. Pollara, and H. van Tilborg. Some easily analyzable convolutional codes. In *Book Abstr. IEEE Intl. Symp. Info. Th.*, page 177, 1990.

[119] A. Drukarev. *Error Control Using Sequential Decoding with Repeat Request.* PhD thesis, I.I.T., Chicago, IL, 1981.

[120] A. Drukarev and D.J. Costello, Jr. ARQ error control using sequential decoding. In *Proc. IEEE Intl. Comm. Conf.*, pages 4.7.1–4.7.5, 1980.

[121] A. Drukarev and D.J. Costello, Jr. A comparison of block and convolutional codes in ARQ error control schemes. *IEEE Trans. Comm.*, COM-30(11):2449–2455, 1982.

[122] A. Drukarev and D.J. Costello, Jr. Hybrid ARQ error control using sequential decoding. *IEEE Trans. Inf. Theory*, IT-29(4):521–535, 1983.

[123] A.I. Drukarev and K.P. Yiu. Performance of error-correcting codes on channels with memory. *IEEE Trans. Comm.*, COM-34(6):513–521, 1986.

[124] J. D'Souza and S.L. Maskara. Simple method for constructing equivalent punctured codes for given true high rate codes. *Elec. Let.*, 30(1):14–16, 1994.

[125] J. Du, M. Kasahara, and T. Namekawa. Seperable codes on type-II hybrid ARQ systems. *IEEE Trans. Comm.*, 36(10):1089–1097, 1988.

[126] P.M. Ebert and S.Y. Tong. Convolutional Reed-Solomon codes. *Bell Sys. Tech. J.*, 48:729–742, 1969.

[127] A. Elahi-Taleghani. *Distance Properties of Rate 1/2 Convolutional Codes.* PhD thesis, I. I. T., Chicago, IL, 1975.

[128] P. Elias. Error-free coding. *IRE Trans. on Info. Th.*, PGIT-4:29–37, 1954.

[129] P. Elias. Coding for noisy channels. In *IRE Conv. Rec., pt. 4*, pages 37–46, 1955.

[130] J. En. A new random-error-correction code. *IEEE Trans. Comp.*, pages 86–88, 1977.

[131] D.D. Falconer. *A Hybrid Sequential and Algebraic Decoding Scheme.* PhD thesis, M. I. T., Cambridge, MA, 1967.

[132] D.D. Falconer. A hybrid coding scheme for discrete memoryless channels. *Bell Sys. Tech. J.*, 48:691–728, 1969.

[133] R. J.F. Fang. Lower bounds on reliability functions of variable-length nonsystematic convolutional codes for channels with noiseless feedback. *IEEE Trans. Inf. Theory*, IT-17(2):161–171, 1971.

[134] R.M. Fano. *Transmission of Information.* John Wiley and Sons, New York, 1961.

[135] R.M. Fano. A heuristic discussion of probabilistic decoding. *IEEE Trans. Inf. Theory*, IT-9:64–74, 1963.

[136] R. Fantacci. Performance evaluation of efficient continuous ARQ protocols. *IEEE Trans. Comm.*, COM-38(6):773–781, 1990.

[137] K. Fazel and P. Salembier. Application of error modeling at the output of maximum likelihood decoder to concatenated coded 16 PSK. In *Proc. IEEE GLOBECOM Conf.*, pages 1528–1533, 1989.

[138] P.M. Feldman. Effect of interference on throughput of adaptive hybrid ARQ protocol. In *Proc. IEEE MILCOM Conf.*, pages 691–696, 1988.

[139] H.C. Ferreira, D.A. Wright, A.S.J. Helberg, I.S. Shaw, and C.R. Wyman. Some new rate $R = k/n$ $(2 \leq k \leq n-2)$ systematic convolutional codes with good distance profiles. *IEEE Trans. Inf. Theory*, IT-37(3):649–653, 1991.

[140] G. Fettweis and H. Meyr. Parallel Viterbi algorithm implementation: Breaking the ACS-bottleneck. *IEEE Trans. Comm.*, 37(8):785–790, 1989.

[141] G. Fettweis and H. Meyr. A 100mbit/s Viterbi decoder chip: Novel architecture and its realisation. In *Proc. IEEE Intl. Comm. Conf.*, pages 463–467, 1990.

[142] G. Fettweis and H. Meyr. High-rate Viterbi processor: A systolic array solution. *IEEE J. Sel. Areas in Comm.*, 8(8):1520–1534, 1990.

[143] G. Fettweis and H. Meyr. High-speed parallel Viterbi decoding: Algorithm and VLSI- architecture. *IEEE Comm. Mag.*, pages 46–55, 1991.

[144] G.D. Forney, Jr. *Concatenated Codes.* M.I.T. Press, Cambridge,MA, 1966.

[145] G.D. Forney, Jr. Exponential error bounds for erasure, list, and decision feedback schemes. *IEEE Trans. Inf. Theory*, IT-14(2):206–220, 1968.

[146] G.D. Forney, Jr. Coding and its applications in space communications. *IEEE Spectrum*, pages 47–58, 1970.

[147] G.D. Forney, Jr. Convolutional codes I: Algebraic structure. *IEEE Trans. Inf. Theory*, IT-16(6):720–738, 1970.

[148] G.D. Forney, Jr. Use of a sequential decoder to analyse convolutional codes structure. *IEEE Trans. Inf. Theory*, IT-:793–795, 1970.

[149] G.D. Forney, Jr. Structural analysis of convolutional codes via dual codes. *IEEE Trans. Inf. Theory*, IT-19(4):512–518, 1973.

[150] G.D. Forney, Jr. The Viterbi algorithm. *Proc. IEEE*, 61(3):268–278, 1973.

[151] G.D. Forney, Jr. Convolutional codes II: Maximum-likelihood decoding. *Inf. and Ctrl.*, 25:222–266, 1974.

[152] G.D. Forney, Jr. Convolutional codes III: Sequential decoding. *Inf. and Ctrl.*, 25:266–297, 1974.

[153] G.D. Forney, Jr. Minimal bases of rational vector spaces, with applications to multivariable linear systems. *SIAM Journal on Control*, 13(3):493–520, 1975.

[154] G.D. Forney, Jr. Algebraic structure of convolutional codes and algebraic system theory. In A.C. Antoulas, editor, *Mathematical System Theory*, pages 527–558. Springer-Verlag, New York, 1991.

[155] G.D. Forney, Jr. Geometrically uniform codes. *IEEE Trans. Inf. Theory*, IT-37(5):1241–1260, 1991.

[156] G.D. Forney, Jr. and E.K. Bower. A high-speed sequential decoder: Prototype design and test. *IEEE Trans. Comm.*, COM-19(5):821–835, 1971.

[157] G.D. Forney, Jr., R.G. Gallager, G.R. Lang, F.M. Longstaff, and S.U. Qureshi. Efficient modulation for band-limited channels. *IEEE J. Sel. Areas in Comm.*, SAC-2(5):632–647, 1984.

[158] G.D. Forney, Jr. and M.D. Trott. The dynamics of group codes: State spaces, trellis diagrams, and canonical encoders. *IEEE Trans. Inf. Theory*, 39(5):1491–1513, 1993.

[159] R.G. Gallager. *Information Theory and Reliable Communication.* John Wiley and Sons, New York, 1968.

[160] J.M. Geist. *Algorithmic Aspects of Sequential Decoding.* PhD thesis, Univ. of Notre Dame, Notre Dame, IN, 1970. Preprint.

[161] J.M. Geist. An empirical comparison of two sequential decoding algorithms. *IEEE Trans. Comm.*, COM-19(4):415–419, 1971.

[162] J.M. Geist and J.B. Cain. Viterbi decoder performance in gaussian noise and periodic erasure bursts. *IEEE Trans. Comm.*, COM-28(8):1417–1422, 1980.

[163] A. Gill. *Linear Sequential Circuits - Analysis, Synthesis, and Applications.* McGraw-Hill, New York, 1966.
[164] M.J.E. Golay. Notes on digital coding. *Proc. I.R.E.*, 37:657, 1949.
[165] R.M. Gray. *Source Coding Theory.* Kluwer Academic Publishers, Boston, 1990.
[166] P.G. Gulak and T. Kailath. Locally connected VLSI architectures for the Viterbi algorithm. *IEEE J. Sel. Areas in Comm.*, 6(3):527–537, 1988.
[167] P.G. Gulak and E. Shwedyk. VLSI structures for Viterbi receivers: Part I - general theory and applications. *IEEE J. Sel. Areas in Comm.*, SAC-4(1):142–154, 1986.
[168] P.G. Gulak and E. Shwedyk. VLSI structures for Viterbi receivers: Part II - encoded MSK modulation. *IEEE J. Sel. Areas in Comm.*, SAC-4(1):155–159, 1986.
[169] D.V. Gupta and J.K. Wolf. A shortened Viterbi decoding algorithm for terminated rate-1/n convolutional codes with hard decisions. *IEEE Trans. Inf. Theory*, IT-25(2):243–246, 1979.
[170] D. Haccoun. Decoding techniques for convolutional codes. In G. Cohen and P. Charpin, editors, *Lec. Notes in Comp. Sci.*, volume 514, pages 242–258. Springer-Verlag, 1990. EUROCODE '90 Proceedings.
[171] D. Haccoun and G. Begin. High-rate punctured convolutional codes for Viterbi and sequential decoding. *IEEE Trans. Comm.*, COM-37(11):1113–1125, 1989.
[172] D. Haccoun and M.J. Ferguson. Generalized stack algorithms for decoding convolutional codes. *IEEE Trans. Inf. Theory*, IT-21(6):638–651, 1975.
[173] D. Haccoun, P. Lavoie, and Y. Savaria. New architectures for fast convolutional encoders and threshold decoders. *IEEE J. Sel. Areas in Comm.*, 6(3):547–557, 1988.
[174] D. Haccoun and P. Montreuil. Algorithms for the spectrum determination of convolutional codes. In *Book Abstr. IEEE Intl. Symp. Info. Th.*, page 49, 1988.
[175] J. Hagenauer. High rate convolutional codes with good distance profiles. *IEEE Trans. Inf. Theory*, IT-23(5):615–618, 1977.
[176] J. Hagenauer. Hybrid ARQ/FEC protocols on fading channels using rate compatible punctured convolutional codes. In *Proc. IEEE Intl. Comm. Conf.*, pages 744–748, 1987.
[177] J. Hagenauer. Rate compatible punctured convolutional codes. In *Proc. IEEE Intl. Comm. Conf.*, pages 1032–1036, 1987.
[178] J. Hagenauer. Rate-compatible punctured convolutional codes (rcpc codes) and their applications. *IEEE Trans. Comm.*, COM-36(4):389–400, 1988.
[179] J. Hagenauer and P. Hoeher. A Viterbi algorithm with soft-decision outputs and its applications. In *Proc. IEEE GLOBECOM Conf.*, pages 1680–1686, 1989.
[180] J. Hagenauer, N. Seshadri, and C.W. Sundberg. The performance of rate-compatible punctured convolutional codes for digital mobile radio. *IEEE Trans. Comm.*, COM-38(7):966–980, 1990.
[181] R.W. Hamming. Error detecting and error correcting codes. *Bell Sys. Tech. J.*, 29(2):147–160, 1950.
[182] M. Hankamer. *Concatenated Convolutional Codes: Application to Deep Space Telecommunications.* PhD thesis, N. M. State Univ., Las Cruces, NM, 1978.
[183] M. Hankamer. Error event statistics for convolutional codes. In *National Teletraffic Conference*, pages 2.6.1–2.6.5, 1978.

[184] P.G. Harrison and N.M. Patel. *Performance Modelling of Communication Networks and Computer Architectures.* Addison-Wesley, Reading, MA, 1993.

[185] B. Hartley and T.O. Hawkes. *Rings, Modules and Linear Algebra.* Chapman and Hall, London, 1970.

[186] R.V. Hartley. Transmission of information. *Bell Sys. Tech. J.*, 7:535–563, 1928.

[187] C R.P. Hartmann and L.D. Rudolph. An optimum symbol-by-symbol decoding rule for linear codes. *IEEE Trans. Inf. Theory*, IT-22(5):514–517, 1976.

[188] B.A. Harvey and S.B. Wicker. Error-trapping Viterbi decoding in type-I hybrid-ARQ protocols. In *Proc. IEEE Intl. Comm. Conf.*, 1990.

[189] B.A. Harvey and S.B. Wicker. Error-trapping Viterbi decoding for type-I hybrid-ARQ protocols. *Can. J. Elec. and Comp. Eng.*, 16(1):5–12, 1991.

[190] T. Hashimoto. A coded ARQ scheme with the generalized Viterbi algorithm. *IEEE Trans. Inf. Theory*, 39(2):423–432, 1993.

[191] S. Haykin. *Digital Communications.* Wiley and Sons, New York, 1988.

[192] H.J. Helgert. Short constraint length rate 1/2 'quick-look' codes. *IEEE Trans. Comm.*, COM-23(7):768–773, 1975.

[193] H.J. Helgert. Correction to 'short constraint length rate 1/2 'quick-look' codes'. *IEEE Trans. Comm.*, COM-24(2):286, 1976.

[194] H.J. Helgert and R.D. Stinaff. Minimum-distance bounds for binary linear codes. *IEEE Trans. Inf. Theory*, IT-19(3):344–356, 1973.

[195] J.A. Heller. Sequential decoding: Short constraint length convolutional codes. Space program summary 37-54, vol. 3, Jet Propulsion Laboratory, California Institute of Technology, Pasadena, CA, 1968.

[196] J.A. Heller. Feedback decoding of convolutional codes. In A.J. Viterbi, editor, *Advances in Communication Systems, vol. 4*, pages 261–278. Academic Press, New York, 1975.

[197] J.A. Heller and I.M. Jacobs. Viterbi decoding for satellite and space communications. *IEEE Trans. Comm.*, COM-19(5):835–848, 1971.

[198] F. Hemmati. *Convolutional Code Structure and the Performance of the Viterbi Algorithm.* PhD thesis, I.I.T., Chicago, IL, 1977.

[199] F. Hemmati and D.J. Costello, Jr. Truncation error probability in Viterbi decoding. *IEEE Trans. Comm.*, pages 530–532, 1977.

[200] F. Hemmati and D.J. Costello, Jr. Asymptotically catastropic convolutional codes. *IEEE Trans. Inf. Theory*, IT-26(3):298–304, 1980.

[201] M.A. Herro, D.J. Costello, Jr., and L. Hu. Capacity abd cutoff rate calculations for a concatenated coding systems. *IEEE Trans. Inf. Theory*, 34(2):212–222, 1988.

[202] M.A. Herro, L. Hu, and J.M. Nowack. Bit error probabillity calculations for convolutional codes with short constraint lenghts on very noisy channels. *IEEE Trans. Comm.*, 36(7):885–888, 1988.

[203] M.G. Hluchyj and M.J. Karol. Queueing in high-performance packet switching. *IEEE J. Sel. Areas in Comm.*, 6(9):1587–1597, 1988.

[204] D.G. Hoffman, D.A. Leonard, C.C. Linder, K.T. Phelps, C.A. Rodger, and J.R. Wall. *Coding Theory - The Essentials.* Marcel Dekker, New York, 1991.

[205] K.J. Hole. New short constraint length rate (n-1)/n punctured convolutional codes for soft-decision Viterbi decoding. *IEEE Trans. Inf. Theory*, IT-34(5):1079–1081, 1988.

[206] K.J. Hole. An algorithm for determining if a rate (n-1)/n punctured convolutional encoder is catastrophic. *IEEE Trans. Comm.*, COM-39(3):386–389, 1991.

[207] K.J. Hole. Punctured convolutional codes for the 1-D partial-response channel. *IEEE Trans. Inf. Theory*, IT-37(3):808–817, 1991.

[208] K.J. Hole. Rate k/(k+1) minimal punctured convolutional encoders. *IEEE Trans. Inf. Theory*, IT-37(3):653–655, 1991.

[209] K.J. Hole and Ø. Ytrehus. Improved coding techniques for precoded partial-response channels. To appear in IEEE Trans. Info. Th.

[210] C.-P. Hsu and C.L. Weber. A soft decision syndrome decoding algorithm for convolutional codes. In *Proc. IEEE MILCOM Conf.*, pages 375–379, 1990.

[211] T.W. Hungerford. *Algebra*. Springer-Verlag, New York, 1989.

[212] G.K. Huth and C.L. Weber. Minimum weight convolutional codewords of finite length. *IEEE Trans. Inf. Theory*, IT-22(2):243–246, 1976.

[213] H. Imai, editor. *Essentials of Error-Control Coding Techniques*. Academic Press, New York, 1990.

[214] T. Ishitani, K. Tansho, N. Miyahara, S. Kubota, and S. Kato. A scarce-state-transition Viterbi-decoder VLSI for bit error correction. *IEEE J. Solid-State Circuits*, 22(4):575–582, 1987.

[215] I.M. Jacobs. Sequential decoding for efficient communication from deep space. *IEEE Trans. Comm.*, COM-15(4):492–501, 1967.

[216] I.M. Jacobs. Practical applications of coding. *IEEE Trans. Inf. Theory*, IT-20(3):305–310, 1974.

[217] F. Jelinek. Sequential decoding algorithm using a stack. *IBM J. Res. and Dev.*, 13:675–678, 1969.

[218] F. Jelinek. Bootstrap trellis decoding. *IEEE Trans. Inf. Theory*, IT-21(3):318–325, 1975.

[219] F. Jelinek and J. Cocke. Bootstrap hybrid decoding for symmetrical binary input channels. *Inf. and Ctrl.*, 18:261–298, 1971.

[220] J.M. Jensen and I.S. Reed. Bounded distance coset decoding of convolutional codes. *IEE Proc.*, 133-F(5):488–492, 1986.

[221] R. Johannesson. Robustly optimal rate one-half binary convolutional codes. *IEEE Trans. Inf. Theory*, IT-21(4):464–468, 1975.

[222] R. Johannesson. Some long rate one-half binary convolutional codes with an optimum distance profile. *IEEE Trans. Inf. Theory*, IT-22(5):629–631, 1976.

[223] R. Johannesson. Some rate 1/3 and 1/4 binary convolutional codes with an optimum distance profile. *IEEE Trans. Inf. Theory*, IT-23(2):281–283, 1977.

[224] R. Johannesson and E. Paaske. Further results on binary convolutional codes with an optimum distance profile. *IEEE Trans. Inf. Theory*, IT-24(2):264–268, 1978.

[225] R. Johannesson and Z.-x. Wan. A linear algebra approach to minimal convolutional encoders. *IEEE Trans. Inf. Theory*, 39(4):1219–1233, 1993.

[226] J. Justesen. New convolutional code constructions and a class of asymptotically good time-varying codes. *IEEE Trans. Inf. Theory*, IT-19(2):220–225, 1973.

[227] J. Justesen. An algebraic construction of rate 1/v convolutional codes. *IEEE Trans. Inf. Theory*, IT-21(5):577–580, 1975.

[228] J. Justesen. Bounded distance decoding of unit memory codes. *IEEE Trans. Inf. Theory*, 39(5):1616–1627, 1993.

[229] J. Justesen and L.R. Hughes. On maximum-distance-seperable convolutional codes. *IEEE Trans. Inf. Theory*, page 288, 1974.

[230] J. Justesen, E. Paaske, and M. Ballan. Quasi-cyclic unit-memory convolutional codes. *IEEE Trans. Inf. Theory*, 36(3):540–547, 1990.

[231] R.E. Kahn, S.A. Gronemeyer, J. Burchfiel, and R.C. Kunzelman. Advances in packet radio technology. *Proc. IEEE*, 66(11):1468–1496, 1978.

[232] T. Kailath. *Linear Systems*. Prentice-Hall, Englewood Cliffs, NJ, 1980.

[233] S. Kallel. Efficient hybrid ARQ protocols with adaptive forward error correction. To appear in IEEE Trans. Comm.

[234] S. Kallel. Analysis of a type II hybrid ARQ scheme with code combining. *IEEE Trans. Comm.*, 38(8):1133–1137, 1990.

[235] S. Kallel. Sequential decoding with an efficient partial retransmission ARQ strategy. *IEEE Trans. Comm.*, COM-39(2):208–213, 1991.

[236] S. Kallel. Efficient stop-and-wait type II hybrid ARQ scheme. *Elec. Let.*, 28(12):1097–1098, 1992.

[237] S. Kallel. Sequential decoding with an efficient incremental redundancy ARQ scheme. *IEEE Trans. Comm.*, COM-40(10):1588–1593, 1992.

[238] S. Kallel. Complementary punctured convolutional (CPC) codes and their use in hybrid ARQ schemes. In *IEEE Pac. Rim Conf. C.C.S.P.*, pages 186–189, 1993.

[239] S. Kallel and D. Haccoun. Sequential decoding with ARQ and code combining. In *Proc. IEEE MILCOM Conf.*, pages 492–496, 1987.

[240] S. Kallel and D. Haccoun. Sequential decoding with ARQ and code combining: A robust hybrid FEC/ARQ system. *IEEE Trans. Comm.*, COM-36(7):773–780, 1988.

[241] S. Kallel and D. Haccoun. Generalised type II hybrid ARQ scheme using punctured convolutional coding. *IEEE Trans. Comm.*, COM-38(11):1938–1946, 1990.

[242] S. Kallel and C. Leung. Type II ARQ schemes with multiple copy decoding for mobile communications. In *Proc. IEEE Veh. Tech. Conf.*, pages 349–353, 1990.

[243] S. Kallel and C. Leung. Adaptive incremental redundancy selective-repeat ARQ scheme for finite buffer receivers. *Elec. Let.*, 28(7):664–666, 1992.

[244] S. Kallel and C. Leung. Efficient ARQ schemes with multiple copy decoding. *IEEE Trans. Comm.*, COM-40(3):642–650, 1992.

[245] L.N. Kanal and A.R.K. Sastry. Models for channels with memory and their applications to error control. *Proc. IEEE*, 66(7):724–744, 1978.

[246] M.J. Karol, M.G. Hluchyj, and S.P. Morgan. Input versus output queueing on a space-division packet switch. *IEEE Trans. Comm.*, COM-35(12):1347–1356, 1987.

[247] T. Kasami, T. Fujiwara, and S. Lin. A concatenated coding scheme for error control. *IEEE Trans. Comm.*, COM-34(5):481–488, 1986.

[248] T. Kitami and I. Tokizawa. Cell loss compensation schemes employing error correction coding for asynchronous broadband ISDN. In *Proc. IEEE INFOCOM Conf.*, pages 116–123, 1990.

[249] A Kohlenberg and G.D. Forney, Jr. Convolutional coding for channels with memory. *IEEE Trans. Inf. Theory*, IT-14(5):618–626, 1968.

[250] H. Koorapaty. Table-driven decoding of binary half rate convolutionally codes with soft decisions. Master's thesis, ECE, NCSU, 1993.

[251] H. Koorapaty, D.L. Bitzer, A. Dholakia, and M.A. Vouk. Table-driven decoding of convolutional codes with soft decision. In *Proc. Joint DIMACS/IEEE Wshp. Coding and Quant.*, 1992.

[252] M.A. Kousa and M. Rahman. An adaptive error control system using hybrid ARQ schemes. *IEEE Trans. Comm.*, COM-39(7):1049–1057, 1991.

[253] H. Krishna and S.J. Morgera. A new error control scheme for ARQ systems. *IEEE Trans. Comm.*, COM-35(10):981–990, 1987.

[254] S. Kubota, S. Honda, M. Morikura, and S. Kato. Concatenated coding scheme employing soft decision for outer codes- dsd(double soft decision) concatenated coding scheme. In *Proc. IEEE Intl. Comm. Conf.*, pages 221–225, 1991.

[255] S. Kubota, K. Ohtani, and S. Kato. High-speed and high-coding-gain Viterbi decoder with low power consumption employing SST (scarce state transistion) scheme. *Elec. Let.*, 22(9):491–493, 1986.

[256] B.D. Kudryashov. Soft decoding for block codes obtained from convolutional codes. In G. Cohen, S. Litsyn, A. Lobstein, and G. Zemer, editors, *Lec. Notes in Comp. Sci.*, volume 573, pages 113–119. Springer-Verlag, 1991. Proceedings of the First French-Soviet Workshop on Algerbaic Coding.

[257] B.D. Kudryashov. Error probability for repeat request systems with convolutional codes. *IEEE Trans. Inf. Theory*, 39(5):1680–1684, 1993.

[258] F.F. Kuo, Ed. *Protocols and Techniques for Data Communications Networks.* Prentice-Hall, Englewood Cliffs, NJ, 1981.

[259] S.K. Leung-Yan-Cheong, E.R. Barnes, and D.U. Friedman. On some properties of the undetected error probability of linear codes. *IEEE Trans. Inf. Theory*, IT-25(1):110–112, 1979.

[260] K.J. Larsen. Comments on "an efficient algorithm for computing free distance. *IEEE Trans. Inf. Theory*, IT-19(4):577–579, 1973.

[261] K.J. Larsen. Short convolutional codes with maximal free distance for rates 1/2, 1/3, and 1/4. *IEEE Trans. Inf. Theory*, IT-19(3):371–372, 1973.

[262] W.W. Larue and V.S. Frost. A technique for extrapolating the end-to-end performance of hdlc links for a range of lost packet rates. *IEEE Trans. Comm.*, COM-38(4):461–466, 1990.

[263] G.S. Lauer. Some optimal partial-unit-memory codes. *IEEE Trans. Inf. Theory*, IT-25(2):240–243, 1979.

[264] P. Lavoie, D. Haccoun, and Y. Savaria. New VLSI architectures for fast soft decision threshold decoders. *IEEE Trans. Comm.*, COM-39(2):200–207, 1991.

[265] J.W. Layland and W.A. Lushbaugh. A flexible high-speed sequential decoder for deep space channels. *IEEE Trans. Comm.*, COM-19(5):813–820, 1971.

[266] E.A. Lee and D.G. Messerschmitt. *Digital Communications.* Kluwer Academic Publishers, Boston, 1988.

[267] L.N. Lee. Real-time minimal-bit-error probability decoding of convolutional codes. *IEEE Trans. Comm.*, COM-22(2):146–151, 1974.

[268] L.N. Lee. *Concatenated Coding Systems Employing Unit-Memory Convolutional Codes and Byte-Oriented Decoding Algorithms.* PhD thesis, Univ. of Notre Dame, Notre Dame, IN, 1976.
[269] L.N. Lee. Short unit-memory byte-oriented binary convolutional codes having maximal free distance. *IEEE Trans. Inf. Theory*, IT-22(3):349–352, 1976.
[270] L.N. Lee. Concatenated coding system employing a unit-memory convolutional code and a byte-oriented decoding algorithm. *IEEE Trans. Comm.*, COM-25(10):1064–1074, 1977.
[271] P.J. Lee. New short constraint length, rate 1/n convolutional codes which minimize the required SNR for given desired bit error rates. *IEEE Trans. Comm.*, COM-33(2):171–177, 1985.
[272] P.J. Lee. Further results on rate 1/n convolutional code constructions with minimum required SNR criterion. *IEEE Trans. Comm.*, COM-34(4):395–399, 1986.
[273] P.J. Lee. Constructions of rate (n-1)/n punctured convolutional codes with minimum required SNR criterion. *IEEE Trans. Comm.*, 36(10):1171–1174, 1988.
[274] D.A. Leonard and Rodger C.A. Limiting error propagation in Viterbi decoding of convolutional codes. *IEEE Trans. Inf. Theory*, IT-35(6):1295–1299, 1989.
[275] K.B. Letaief and J.S. Sadowsky. New importance sampling methods for simulating sequential decoders. *IEEE Trans. Inf. Theory*, 39(5):1716–1722, 1993.
[276] C.S.K. Leung and A. Lam. Forward error correction for an ARQ scheme. *IEEE Trans. Comm.*, COM-29(10):1514–1519, 1981.
[277] G. Li, Q. Wang, V.K. Bhargava, and L.J. Mason. Repeated convolutional codes for high error rate channel. In *Proc. IEEE MILCOM Conf.*, pages 135–140, 1990.
[278] K. Li and S. Kallel. Bidirectional sequential decoding for convolutional codes. In *IEEE Pac. Rim Conf. C.C.S.P.*, pages 200–205, 1991.
[279] R. Lidl and H. Niederreiter. *Finite Fields*, volume 20 of *Encyclopedia of Mathematics and Its Applications.* Addison-Wesley, Reading, MA, 1983.
[280] C.-F. Lin and J.B. Anderson. M-algorithm decoding of channel convolutional codes. In *Proc. Conf. Info. Sci. Sys.*, pages 362–366, Princeton, NJ, 1986.
[281] C.-F. Lin and J.B. Anderson. M-algorithm decoding with path recovery for convolutional channel codes. In *Proc. IEEE GLOBECOM Conf.*, pages 181–185, 1986.
[282] H.-D. Lin and D.G. Messerschmitt. Algorithms and architecture for concurrent Viterbi decoding. In *Proc. IEEE Intl. Comm. Conf.*, pages 836–840, 1989.
[283] M.-C. Lin and M.-Y. Guu. The performance analysis of a concatenated ARQ scheme using parity retransmissions. *IEEE Trans. Comm.*, COM-39(12):1869–1874, 1991.
[284] S. Lin. *An Introduction to Error Correcting Codes.* Prentice-Hall, Englewood Cliffs, New Jersey, 1970.
[285] S. Lin and D.J. Costello, Jr. *Error Control Coding - Fundamentals and Applications.* Prentice-Hall, Englewood Cliffs,NJ 07632, 1983.
[286] S. Lin, D.J. Costello, Jr., and M.J. Miller. Automatic-repeat-request error-control schemes. *IEEE Comm. Mag.*, 1984.
[287] S. Lin and H. Lyne. Some results on binary convolutional code generators. *IEEE Trans. Inf. Theory*, IT-13(1):134–139, 1967.

[288] S. Lin and Y.-M. Wang. A parity retransmission hybrid ARQ using a convolutional code and Viterbi decoding for error control. In *Proc. IEEE GLOBECOM Conf.*, pages 1097–1100, 1982.

[289] S. Lin and P.S. Yu. An effective error-control scheme for satellite communications. *IEEE Trans. Comm.*, COM-28(3):395–401, 1980.

[290] S. Lin and P.S. Yu. A hybrid ARQ scheme with parity retransmission for error control of satellite channels. In *Proc. IEEE Natl. Telecomm. Conf.*, pages G10.3.1–G10.3.8, 1981.

[291] S. Lin and P.S. Yu. A hybrid ARQ scheme with parity retransmission for error control of satellite channels. *IEEE Trans. Comm.*, COM-30(7):1701–1719, 1982.

[292] K.Y. Liu and J.-J. Lee. Recent results on the use of concatenated Reed-Solomon/Viterbi channel coding and data compression for space communications. *IEEE Trans. Comm.*, COM-32(5):518–523, 1984.

[293] H.-A. Loeliger. *On Euclidean-Space Group Codes.* PhD thesis, Swiss Federal Institute of Technology, Zurich, Switzerland, 1992.

[294] G. Longo, Ed. *Coding and Complexity: CISM Courses and Lectures no. 216.* Springer Verlag, New York, 1975.

[295] H. Lou and J. M. Cioffi. A programmable parallel processor architecture for Viterbi detection. In *Proc. IEEE GLOBECOM Conf.*, pages 1333–1337, 1990.

[296] L. Lugand. *ARQ Schemes Using Convolutional Codes and Viterbi Decoding Over Non-stationary Channels.* PhD thesis, I.I.T., Chicago, IL, 1981.

[297] L. Lugand and D.J. Cotello, Jr. A comparison of three hybrid ARQ schemes using convolutional codes on a non-stationary channel. In *Proc. IEEE GLOBECOM Conf.*, pages 702–706, 1982.

[298] L.R. Lugand, D.J. Costello, Jr., and R.H. Deng. Parity retransmission hybrid ARQ using rate 1/2 convolutional codes on a nonstationary channel. *IEEE Trans. Comm.*, COM-37(7):755–765, 1989.

[299] H.H. Ma. The multiple stack algorithm implemented on a zilog z-80 microcomputer. *IEEE Trans. Comm.*, COM-28(11):1876–1882, 1980.

[300] H.H. Ma. *Generalized tail biting convolutional codes.* PhD thesis, University of Massachusetts, Amherst, MA, 1985.

[301] H.H. Ma and J.K. Wolf. Binary unequal error-protection block codes formed from convolutional codes by generalized tail-biting. *IEEE Trans. Inf. Theory*, IT-32(6):776–786, 1986.

[302] H.H. Ma and J.K. Wolf. On tail biting convolutional codes. *IEEE Trans. Comm.*, COM-34(2):104–111, 1986.

[303] F.J. MacWilliams and N.J.A. Sloane. *The Theory of Error-Correcting Codes: Part I and II.* North-Holland, New York, 1977.

[304] D.M. Mandelbaum. An adaptive-feedback coding scheme using incremental redundancy. *IEEE Trans. Inf. Theory*, IT-20(3):388–389, 1974.

[305] D.M. Mandelbaum. On forward error correction with adaptive decoding. *IEEE Trans. Inf. Theory*, IT-21(2):230–233, 1975.

[306] I. Markman. *Sequential Decoding on Channels with and without Memory Under a Probability of Error Criterion.* PhD thesis, RPI., Troy, NY, 1993.

[307] J.A.C. Martins and J. de C. Alves. ARQ protocols with adaptive block size perform better over a wide range of bit error rates. *IEEE Trans. Comm.*, COM-38(6):737–739, 1990.

[308] J.L. Massey. *Threshold Decoding.* M.I.T. Press, Cambridge,MA, 1963.

[309] J.L. Massey. Uniform codes. *IEEE Trans. Inf. Theory*, IT-12(2):132–134, 1966.

[310] J.L. Massey. Catastrophic error-propagation in convolutional codes. In *Proc. 11th Midwest Symp. Cir. Th.*, pages 583–587, Notre Dame, IN, 1968.

[311] J.L. Massey. Some algebraic and distance properties of convolutional codes. In H.B. Mann, editor, *Error Correcting Codes; Proceedings of a symposium*, pages 89–109. Wiley, New York, 1968.

[312] J.L. Massey. Variable-length codes and the Fano metric. *IEEE Trans. Inf. Theory*, IT-18(1):196–198, 1972.

[313] J.L. Massey. Coding and modulation in digital communications. In *Proc. IEEE Intl. Zurich Sem. Dig. Comm.*, pages 248–250, Zurich, Switzerland, 1974.

[314] J.L. Massey. Advances in threshold decoding. In A.V. Balakrishnan, editor, *Advances in Communication Systems, vol. 3.* Academic Press, New York, 1975.

[315] J.L. Massey. Error bounds for tree codes, trellis codes and convolutional codes with encoding and decoding procedures. In G. Longo, editor, *Coding and Complexity: CISM No. 216*, pages 1–58. Springer-Verlag, New York, 1975.

[316] J.L. Massey. Information theory : The copernican system of communications. *IEEE Comm. Mag.*, 22(12):26–28, 1984.

[317] J.L. Massey. Coding theory. In W. Ledermann and S. Vajda, editors, *Handbook of Applicable Mathematics: Vol. V, Part B*, chapter 16, pages 623–676. Wiley, New York, 1985.

[318] J.L. Massey and D.J. Costello, Jr. Nonsystematic convolutional codes for sequential decoding in space applications. *IEEE Trans. Comm.*, COM-19(5):806–813, 1971.

[319] J.L. Massey, D.J. Costello, Jr., and J. Justesen. Polynomial weights and code constructions. *IEEE Trans. Inf. Theory*, IT-19(1):101–110, 1973.

[320] J.L. Massey and R.W. Liu. Application of lyapunov's direct method to the error-propagation effect in convolutional codes. *IEEE Trans. Inf. Theory*, IT-10:248–250, 1964.

[321] J.L. Massey and M.K. Sain. Codes, automata, and continuous systems: Explicit interconnections. *IEEE Trans. Aut. Cont.*, AC-12:644–650, 1967.

[322] J.L. Massey and M.K. Sain. Inverses of linear sequential circuits. *IEEE Trans. Comp.*, C-17:330–337, 1968.

[323] A.J. McAuley. Reliable broadband communication using a burst erasure correcting code. In *Proc. ACM SIGCOMM*, 1990.

[324] R.J. McEliece. *The Theory of Information and Coding*, volume 3 of *Encyclopedia of Mathematics and Its Applications.* Addison-Wesley, Reading, MA, 1977.

[325] R.J. McEliece. *Finite Fields for Computer Scientists and Engineers.* Kluwer Academic Publishers, Boston, MA, 1987.

[326] R.J. McEliece and I. Onyszchuk. The extended invariant factor algorithm with application to the Forney analysis of convolutional codes. In *Proc. IEEE Intl. Symp. Info. Th.*, page 142, 1993.

[327] J.J. Metzner. Improvements in block-retransmission schemes. *IEEE Trans. Comm.*, COM-27(2):524–532, 1979.

[328] J.J. Metzner and D. Chang. Efficient selective repeat ARQ strategies for very noisy channels. In *Proc. IEEE GLOBECOM Conf.*, pages 1207–1214, 1983.

[329] A. Michelson and G. Rosen. A severely punctured convolutional code: Its performance and structure. In *Proc. IEEE MILCOM Conf.*, pages 1007–1012, 1991.

[330] A.M. Michelson and A.M. Levesque. *Error-Control Techniques for Digital Communication*. John Wiley and Sons, New York, 1985.

[331] A. Miczo and L.D. Rudolph. A note on the free distance of a convolutional code. *IEEE Trans. Inf. Theory*, IT-16(5):646–648, 1970.

[332] M.J. Miller. *Automatic-Repeat-Request Systems for Error Control in Digital Transmission*. PhD thesis, University of Hawaii, Honolulu, HI, 1982.

[333] M.J. Miller and S. Lin. The analysis of some selective-repeat ARQ schemes with finite receiver buffer. *IEEE Trans. Comm.*, COM-29(9):1307–1315, 1981.

[334] S.P. Miller, N. Becker, and P.N. Johnson. Custom ASIC development for high-speed Viterbi decoding. In *Proc. IEEE MILCOM Conf.*, pages 239–243, 1990.

[335] K. Mokrani and S.S. Soliman. Concatenated codes over fading dispersive channels. In *Proc. IEEE Intl. Comm. Conf.*, pages 1378–1382, 1989.

[336] M.-J. Montpetit, D. Haccoun, and G. Deslauriers. A branching process analysis of the stack algorithms for variable channel conditions. *IEEE Trans. Inf. Theory*, 38(6):1845–1848, 1992.

[337] B.C. Moore and L.M. Silvermann. A new characterization of feedforward delay-free inverses. *IEEE Trans. Inf. Theory*, pages 126–129, 1973.

[338] S.D. Morgera and V.K.Oduol. Soft-decision decoding applied to the generalized type II hybrid ARQ scheme. *IEEE Trans. Comm.*, COM-37(4):393–396, 1989.

[339] T.N. Morrissey, Jr. Analysis of decoders for convolutional codes by stochastic sequential machine methods. *IEEE Trans. Inf. Theory*, IT-16(4):460–469, 1970.

[340] W.-H. Ng and P.E. Pfeiffer. Minimum-Hamming-distance decoding of single-generator binary convolutional codes. *Inf. and Ctrl.*, 13:295–315, 1968.

[341] T. Niinomi, I. Sasase, and S. Mori. Selective repeat type II hybrid ARQ/FEC scheme using rate-compatible punctured convolutional code. In *Proc. IEEE Intl. Comm. Conf.*, pages 1251–1255, 1990.

[342] A.A. Nilsson and F.-Y. Lai. Performance evaluation of error recovery schemes in high speed networks. In *Proc. IEEE Intl. Comm. Conf.*, pages 722–726, 1990.

[343] H. Nussbaumer. *Data Circuits, Error Detection, Data Links*, volume 1 of *Computer Communication Systems*. John Wiley and Sons, New York, 1990.

[344] H. Nyquist. Certain factors affecting telegraph speed. *Bell Sys. Tech. J.*, 3:324–346, 1924.

[345] J.P. Odenwalder. *Optimal Decoding of Convolutional Codes*. PhD thesis, UCLA, 1970.

[346] V.K. Oduol and S.D. Morgera. Performance evaluation of the generalized type-II hybrid ARQ scheme with noisy feedback on markov channels. *IEEE Trans. Comm.*, 41(1):32–40, 1993.

[347] H. Ohta and T. Kitami. A cell loss recovery method using FEC in ATM networks. *IEEE J. Sel. Areas in Comm.*, 9(9):1471–1483, 1991.

[348] H. Ohta and T. Kitami. A technique to detect and compensate consecutive cell loss in ATM networks. In *Proc. IEEE INFOCOM Conf.*, pages 781–790, 1991.

[349] R.R. Olson. Note on feedforward inverses for linear sequential circuits. *IEEE Trans. Comp.*, C-19:1216–1221, 1970.

[350] J.K. Omura. On the Viterbi algorithm. *IEEE Trans. Inf. Theory*, IT-15(1):177–179, 1969.

[351] I.M. Onyszchuk. Truncation length for Viterbi decoding. *IEEE Trans. Comm.*, COM-39(7):1023–1026, 1991.

[352] I.M. Onyszchuk, K.-M. Cheung, and O. Collins. Quantization loss in convolutional decoding. *IEEE Trans. Comm.*, 41(2):261–265, 1993.

[353] H. Oshtoff. *Reduced Complexity Decoding of Systematic Encoders.* PhD thesis, University of Lund, Lund, Sweden, 1993.

[354] E. Paaske. Short binary convolutional codes with maximal free distance for rates 2/3 and 3/4. *IEEE Trans. Inf. Theory*, pages 683–689, 1974.

[355] E. Paaske. Comments on 'a new random-error-correction code. *IEEE Trans. Comp.*, C-28(3):255–258, 1979.

[356] R. Palazzo, Jr. A time-varying convolutional encoder better than the best time-invariant encoder. *IEEE Trans. Inf. Theory*, 39(3):1109–1110, 1993.

[357] K.E. Perry and J.M. Wozencraft. SECO: A self-regulating error correcting coder-decoder. *IRE Trans. Inf. Theory*, IT-8:128–135, 1962.

[358] W. W. Peterson and E.J. Weldon, Jr. *Error-Correcting Codes.* M.I.T. Press, Cambridge, MA, 1961.

[359] Ph. Piret. Structure and constructions of cyclic convolutional codes. *IEEE Trans. Inf. Theory*, IT-22(2):147–155, 1976.

[360] Ph. Piret. *Algebraic Properties of Convolutional Codes with Automorphisms.* PhD thesis, Universite Catholique de Louvain, France, 1977.

[361] Ph. Piret. Multiple-word correcting convolutional codes. *IEEE Trans. Inf. Theory*, IT-30(4):637–644, 1984.

[362] Ph. Piret. *Convolutional Codes: An Algebraic Approach.* M.I.T. Press, Cambridge, MA, 1988.

[363] Ph. Piret. A conjecture on free distance of (2,1,m) binary convolutional codes. In L. Huguet, editor, *Lec. Notes in Comp. Sci.*, volume 356, pages 369–375. Springer-Verlag, 1989. AAECC-5 Proceedings.

[364] K.A. Post. Explicit evaluation of Viterbi's union bounds on convolutional code performance for the binary symmetric channel. *IEEE Trans. Inf. Theory*, IT-23(3):403–404, 1977.

[365] J.G. Proakis. *Digital Communications.* McGraw-Hill, New York, 1989.

[366] M. De Prycker. *Asynchronous Transfer Mode : Solutions for Broadband ISDN.* Ellis Horwood, New York, 1991.

[367] Qualcomm. *Q1650 K=7 Multi-Code Rate Viterbi decoder technical data sheet*, 1990.

[368] C.M. Rader. Memory management in a Viterbi decoder. *IEEE Trans. Comm.*, COM-29(9):1399–1401, 1981.

[369] I.S. Reed and T.K. Truong. New syndrome decoder for (n, 1) convolutional codes. *Elec. Let.*, 19(9):344–346, 1983.

[370] I.S. Reed and T.K. Truong. New syndrome decoding techniques for the (n, k) convolutional codes. *IEE Proc.*, 131-F(4):412–416, 1984.

[371] I.S. Reed and T.K. Truong. Error-trellis syndrome decoding techniques for convolutional codes. *IEE Proc.*, 132-F(2):77–83, 1985.

[372] T.C. Reiner and M.J. Lindsey. VLSI development of a reconfigurable multi-user Viterbi decoder. In *Proc. IEEE MILCOM Conf.*, pages 244–248, 1990.

[373] M.Y. Rhee. *Error Correcting Coding Theory.* McGraw-Hill, New York, 1989.

[374] M.D. Rice and S.B. Wicker. Majority logic decoding in type-I hybrid-ARQ protocols. In *Proc. IEEE Intl. Comm. Conf.*, 1990.

[375] M.D. Rice and S.B. Wicker. Modified majority logic decoding of cyclic codes in hybrid-arq systems. *IEEE Trans. Comm.*, 40(9):1413–1417, 1992.

[376] M.P. Ristenbatt. Alternatives in digital communications. *Proc. IEEE*, 61(6):703–721, 1973.

[377] J. P. Robinson. Error propagation and definite decoding of convolutional codes. *IEEE Trans. Inf. Theory*, IT-14(1):121–128, 1968.

[378] J.P. Robinson. An upper bound on the minimum distance of a convolutional code. *IEEE Trans. Inf. Theory*, IT-11(4):567–571, 1965.

[379] J.P. Robinson. Punctured uniform codes. *IEEE Trans. Inf. Theory*, IT-15(1):149–152, 1969.

[380] J.P. Robinson and A.J. Bernstein. A class of binary recurrent codes with limited error propagation. *IEEE Trans. Inf. Theory*, IT-13(1):106–113, 1967.

[381] E.Y. Rocher and R.L. Pickholtz. An analysis of the effectiveness of hybrid transmission schemes. *IBM J. Res. and Dev.*, 14:426–433, 1970.

[382] W.E. Rodgers and R.B. Lackey. Burst-error-correcting convolutional codes with short constraint length. *IEEE Trans. Inf. Theory*, IT-26(3):354–359, 1980.

[383] S. Roman. *Advanced Linear Algebra.* Springer-Verlag, New York, 1992.

[384] S. Roman. *Coding and Information Theory.* Springer-Verlag, New York, 1992.

[385] C. Roos. On the structure of convolutional and cyclic convolutional codes. *IEEE Trans. Inf. Theory*, IT-25(6):676–683, 1979.

[386] W.J. Rosenberg. *Structural Properties of Convolutional Codes.* PhD thesis, UCLA, 1971.

[387] W.J. Rosenberg. Consecutive zeros in convolutional codewords. *Discrete Mathematics*, 3:247–264, 1972.

[388] M. Rouanne and D.J. Costello, Jr. An algorithm for computing the distance spectrum of trellis codes. *IEEE J. Sel. Areas in Comm.*, 7(6):929–940, 1989.

[389] L.D. Rudolph. Generalized threshold decoding of convolutional codes. *IEEE Trans. Inf. Theory*, IT-16(6):739–745, 1970.

[390] P.R. Sadot. Realization of a monochip Viterbi decoder in ASIC technology. In *Proc. IEEE MILCOM Conf.*, pages 249–253, 1990.

[391] A. Said and R. Palazzo, Jr. Using combinatorial optimization to design good unit-memory convolutional codes. *IEEE Trans. Inf. Theory*, 39(3):1100–1108, 1993.

[392] M.K. Sain and J.L. Massey. Invertibility of linear time-invariant dynamical systems. *IEEE Trans. Aut. Cont.*, AC-14:141–149, 1969.

[393] A.R.K. Sastry. Improving automatic repeat-request (arq) performance on satellite channels under high error rate conditions. *IEEE Trans. Comm.*, COM-23(4):436–439, 1975.

[394] A.R.K. Sastry. Performance of hybrid error control schemes on satellite channels. *IEEE Trans. Comm.*, COM-23(7):689–694, 1975.

[395] T. Sato, M. Kawabe, T. Kato, and A. Fukasawa. Throughput analysis method for hybrid ARQ schemes over burst error channels. *IEEE Trans. Veh. Tech.*, 42(1):110–118, 1993.

[396] J.E. Savage. Sequential decoding- the computation problem. *Bell Sys. Tech. J.*, pages 148–175, 1966.

[397] J.E. Savage. Advances in threshold decoding. In A.V. Balakrishnan, editor, *Advances in Communication Systems, vol. 3*. Academic Press, New York, 1975.

[398] J.P.M. Schalkwijk. A coding scheme for additive noise channels with feedback - part II: Band-limited signals. *IEEE Trans. Inf. Theory*, IT-12(2):183–189, 1966.

[399] J.P.M. Schalkwijk and T. Kailath. A coding scheme for additive noise channels with feedback - part I: No bandwidth constraint. *IEEE Trans. Inf. Theory*, IT-12(2):172–182, 1966.

[400] J.P.M. Schalkwijk, K.A. Post, and J.P.J.C. Aarts. On a method of calculating the event error probability of convolutional codes with maximum likelihood decoding. *IEEE Trans. Inf. Theory*, IT-25(6):737–743, 1979.

[401] J.P.M. Schalkwijk and A.J. Vinck. Syndrome decoding of convolutional codes. *IEEE Trans. Comm.*, COM-23(7):789–792, 1975.

[402] J.P.M. Schalkwijk and A.J. Vinck. Syndrome decoding of binary rate 1/2 convolutional codes. *IEEE Trans. Comm.*, COM-24(9):977–985, 1976.

[403] J.P.M. Schalkwijk, A.J. Vinck, and K.A. Post. Syndrome decoding of binary rate k/n convolutional codes. *IEEE Trans. Inf. Theory*, IT-24(5):553–562, 1978.

[404] C.B. Schlegel and M. A. Herro. A burst-error-correcting Viterbi algorithm. *IEEE Trans. Comm.*, COM-38(3):285–291, 1990.

[405] R. Schweikert and A.J. Vinck. A convolutional single-parity-check concatenated coding scheme for high-data-rate applications. *IEEE Trans. Comm.*, 39(1):4–7, 1991.

[406] G. Seguin. On a class of convolutional codes. *IEEE Trans. Inf. Theory*, IT-29(2):215–223, 1983.

[407] N. Seshadri and C.-E.W. Sundberg. Generalized Viterbi algorithms for error detection with convolutional codes. In *Proc. IEEE GLOBECOM Conf.*, pages 1534–1538, 1989.

[408] R. Sfez and G. Battail. A weighted-output variant of the Viterbi algorithm for concatenated schemes using a convolutional inner code. In G. Cohen and P. Charpin, editors, *Lec. Notes in Comp. Sci.*, volume 514, pages 259–270. Springer-Verlag, 1990. EUROCODE '90 Proceedings.

[409] N. Shacham. Performance of ARQ with sequential decoding over one-hop and two-hop radio link. *IEEE Trans. Comm.*, COM-31(10):1172–1180, 1983.

[410] N. Shacham. ARQ with sequential decoding of packetized data: Queueing analysis. *IEEE Trans. Comm.*, COM-32(10):1118–1127, 1984.

[411] N. Shacham. Packet recovery and error correction in high-speed wide-area networks. In *Proc. IEEE MILCOM Conf.*, pages 551–557, 1989.

[412] N. Shacham and P. McKenney. Packet recovery in high-speed networks using coding and buffer management. In *Proc. IEEE INFOCOM Conf.*, pages 124–131, 1990.

[413] C.E. Shannon. A mathematical theory of communication. *Bell Sys. Tech. J.*, 27:379–423, 623–656, 1948.

[414] C.E. Shannon, R.G. Gallager, and E.R. Berlekamp. Lower bounds to error probabiliy for coding on discrete memoryless channels.1. *Inf. and Ctrl.*, 10:65–103, 1967.

[415] C.E. Shannon, R.G. Gallager, and E.R. Berlekamp. Lower bounds to error probabiliy for coding on discrete memoryless channels.2. *Inf. and Ctrl.*, 10:522–552, 1967.

[416] C.E. Shannon and W. Weaver. *The Mathematical Theory of Communication.* University of Illinois Press, Urbana, IL, 1949.

[417] K.A. Shridhara. A study of the bit error performance of convolutional coded spread spectrum communication systems. *Journal of IETE*, 35(3):160–169, 1989.

[418] C.B. Shung, H.-D. Lin, R. Cypher, P.H. Siegel, and H.K. Thapar. Area-efficient architectures for the Viterbi algorithm - part I: Theory. *IEEE Trans. Comm.*, 41(4):636–644, 1993.

[419] C.B. Shung, H.-D. Lin, R. Cypher, P.H. Siegel, and H.K. Thapar. Area-efficient architectures for the Viterbi algorithm - part II: Applications. *IEEE Trans. Comm.*, 41(5):802–807, 1993.

[420] T.J. Shusta. *Minimal Convolutional Encoders.* PhD thesis, I. I. T., Chicago, IL, 1973.

[421] T.J. Shusta. Enumeration of minimal convolutional encoders. *IEEE Trans. Inf. Theory*, IT-23(1):127–132, 1977.

[422] P.S. Sindhu. Retransmission error control with memory. *IEEE Trans. Comm.*, COM-25(5):473–479, 1977.

[423] B. Sklar. A structured overview of digital communications - a tutorial review - part I. *IEEE Comm. Mag.*, 21(8):4–17, 1983.

[424] B. Sklar. A structured overview of digital communications - a tutorial review - part II. *IEEE Comm. Mag.*, 21(10):4–21, 1983.

[425] B. Sklar. *Digital Communications: Fundamentals and Applications.* Prentice-Hall, Englewood Cliffs, NJ, 1988.

[426] B. Sklar. A tutorial on convolutional coding for m-ary signals - trellis-coded modulation. In *Proc. IEEE MILCOM Conf.*, pages 637–645, El Segundo CA, 1988.

[427] N.J.A. Sloane and A.D. Wyner, editors. *Claude Elwood Shannon : Collected Papers.* IEEE Press, New York, 1993.

[428] J. Snyders. Partial ordering of error patterns for maximum likelihood soft decoding. In G. Cohen, S. Litsyn, A. Lobstein, and G. Zemer, editors, *Lec. Notes in Comp. Sci.*, volume 573, pages 120–125. Springer-Verlag, 1991. Proceedings of the First French-Soviet Workshop on Algerbaic Coding.

[429] G. Solomon and H.C.A. van Tilborg. A connection between block and convolutional codes. *SIAM J. Appl. Math.*, 37(2):358–369, 1979.

[430] W. Stallings. *Data and Computer Communications.* MacMillan, New York, 1988.

[431] I.G. Stiglitz. *Sequential Decoding with Feedback.* PhD thesis, M.I.T., Lexington, MA, 1963.

[432] D.D. Sullivan. *Control of Error Propagation in Convolutional Codes.* PhD thesis, Univ. of Notre Dame, Notre Dame, IN, 1966.

[433] D.D. Sullivan. Implementation of single error correcting convolutional codes. In *Proc. 11th Midwest Symp. Cir. Th.*, pages 588–597, Notre Dame, IN, 1968.

[434] D.D. Sullivan. Error-propagation properties of uniform codes. *IEEE Trans. Inf. Theory*, IT-15(1):152–161, 1969.

[435] J. Sun, I.S. Reed, H.E. Huey, and T.K. Truong. Pruned-trellis search techniques for high-rate convolutional codes. *IEE Proc.*, 136-E(5):405–414, 1989.

[436] P. Sweeney. *Error Control Coding - An Introduction.* Prentice Hall, New York, 1991.

[437] H. Tanaka, K. Furusawa, and S. Kaneku. A novel approach to soft decision decoding of threshold decodable codes. *IEEE Trans. Inf. Theory*, IT-26(2):244–246, 1980.

[438] H.K. Thapar and J.M. Cioffi. A block processing method for designing high-speed Viterbi detectors. In *Proc. IEEE Intl. Comm. Conf.*, pages 1096–1100, 1989.

[439] C. Thommesen and J. Justesen. Bounds on distances and error exponents of unit memory codes. *IEEE Trans. Inf. Theory*, IT-29(5):637–649, 1983.

[440] D. Towsley and J.K. Wolf. On the statistical analysis of queue lengths and waiting times for statistical multiplexers with ARQ retransmission schemes. *IEEE Trans. Comm.*, COM-27(4):693–701, 1979.

[441] K.-H. Tzou and J.G. Dunham. Sliding block decoding of convolutional codes. *IEEE Trans. Comm.*, COM-29(9):1401–1403, 1981.

[442] G. Underboeck. Channel coding with multilevel/phase signals. *IEEE Trans. Inf. Theory*, IT-28(1):55–67, 1982.

[443] G. Ungerboeck. Trellis-coded modulation with redundant signal sets, part I: Introduction. *IEEE Comm. Mag.*, 25(2):5–11, 1987.

[444] G. Ungerboeck. Trellis-coded modulation with redundant signal sets, part II: State of the art. *IEEE Comm. Mag.*, 25(2):12–21, 1987.

[445] G. Ungerboeck and I. Csajka. On improving data-link performance by increasing channel alphabet and introducing sequence coding. In *Book Abstr. IEEE Intl. Symp. Info. Th.*, 1976.

[446] L. van De Meeberg. A tightened upper bound on the error probability of binary convolutional codes with Viterbi decoding. *IEEE Trans. Inf. Theory*, pages 389–391, 1974.

[447] J.H. van Lint. *Introduction to Coding Theory.* Springer-Verlag, New York, 1992.

[448] S.A. Vanstone and P.C. van Oorschot. *An Introduction to Error Correcting Codes with Applications.* Kluwer Academic Publishers, Boston, MA, 1989.

[449] A.J. Vinck, A.J.P. de Paepe, and J.P.M. Schalkwijk. A class of binary rate one-half convolutional codes that allows an improved stack decoder. *IEEE Trans. Inf. Theory*, IT-26(4):389–392, 1980.

[450] A.J. Viterbi. *Principles of Coherent Communication.* McGraw-Hill, New York, 1966.

[451] A.J. Viterbi. Error bounds for convolutional codes and an asymptotically optimum decoding algorithm. *IEEE Trans. Inf. Theory*, IT-13(2):260–269, 1967.

[452] A.J. Viterbi. Convolutional codes and their performance in communication systems. *IEEE Trans. Comm.*, COM-19(5):751–771, 1971.

[453] A.J. Viterbi and J.P. Odenwalder. Further results on optimal decoding of convolutional codes. *IEEE Trans. Inf. Theory*, IT-:732–734, 1969.

[454] A.J. Viterbi and J.K. Omura. *Principles of Digital Communication and Coding.* McGraw-Hill, New York, 1979.

[455] A.J. Viterbi, Ed. *Advances in Communication Systems, vol. 4.* Academic Press, New York, 1975.

[456] T.J. Wagner. A gilbert bound for periodic binary convolutional codes. *IEEE Trans. Inf. Theory*, IT-14(5):752–755, 1968.

[457] Q. Wang and V.K. Bhargava. An efficient maximum likelihood decoding algorithm for generalized tail biting convolutional codes including quasicyclic codes. *IEEE Trans. Comm.*, 37(8):875–879, 1989.

[458] Q. Wang, V.K. Bhargava, and A.T. Gulliver. Construction of optimal quasicyclic codes based on optimum distance profile convolutional codes. In *IEEE Pac. Rim Conf. C.C.S.P.*, pages 105–108, 1987.

[459] Y.-M. Wang and S. Lin. A modified selective-repeat type-II hybrid ARQ system. In *Proc. IEEE Natl. Telecomm. Conf.*, pages G10.4.1–G10.4.5, 1981.

[460] Y.-M. Wang and S. Lin. A modified selective-repeat type-II hybrid ARQ system and its performance analysis. *IEEE Trans. Comm.*, COM-31(5):593–608, 1983.

[461] E.J. Weldon, Jr. An improved selective-repeat ARQ strategy. *IEEE Trans. Comm.*, COM-30(3):480–486, 1982.

[462] K.-A. Wen and J.-F. Wang. A new transform algorithm for Viterbi decoding. *IEEE Trans. Comm.*, COM-38(6):764–772, 1990.

[463] S.B. Wicker. An adaptive type-I hybrid-ARQ technique using the Viterbi decoding algorithm. In *Proc. IEEE MILCOM Conf.*, 1988.

[464] S.B. Wicker. Adaptive rate error control through the use of diversity combining and majority logic decoding in a hybrid-ARQ protocol. In *Proc. IEEE MILCOM Conf.*, 1989.

[465] S.B. Wicker. Hybrid-ARQ Reed-Solomon coding in an adaptive rate system. In *Proc. IEEE Intl. Comm. Conf.*, 1989.

[466] S.B. Wicker. Modified majority-logic decoders for use in convolutionally encoded hybrid-ARQ systems. *IEEE Trans. Comm.*, COM-38(3):263–266, 1990.

[467] S.B. Wicker. Adaptive rate error control through the use of diversity combining and majority logic decoding in a hybrid-ARQ protocol. *IEEE Trans. Comm.*, COM-39(3):380–385, 1991.

[468] S.B. Wicker. Type-II hybrid-ARQ protocols using punctured Reed-Solomon codes. In *Proc. IEEE MILCOM Conf.*, pages 1229–1234, 1991.

[469] S.B. Wicker. *Error Control Systems for Digital Communication and Storage.* Prentice-Hall, Englewood Cliffs, NJ, Summer, 1994.

[470] D. Wiggert. *Codes for Error Control and Synchronization.* Artech House, Norwood, MA, 1988.

[471] J.K. Wolf. A survey of coding theory: 1967-1972. *IEEE Trans. Inf. Theory*, IT-19(4):381–389, 1973.

[472] J.K. Wolf, A.M. Michelson, and A.H. Levesque. On the probability of undetected error for linear block codes. *IEEE Trans. Comm.*, COM-30(2):317–324, 1982.

[473] J. M. Wozencraft and I.M. Jacobs. *Principles of Communication Engineering.* John Wiley and Sons, New York, 1965.

[474] J. M. Wozencraft and B. Rieffen. *Sequential Decoding.* M.I.T. Press and Wiley, Cambridge, MA and New York, 1961.

[475] J.M. Wozencraft. *Sequential Decoding for Reliable Communication.* PhD thesis, M.I.T., Cambridge, MA, 1957.

[476] W.W. Wu. New convolutional codes - part I. *IEEE Trans. Comm.*, COM-23(9):942–956, 1975.

[477] W.W. Wu. New convolutional codes - part II. *IEEE Trans. Comm.*, COM-24(1):19–33, 1976.
[478] W.W. Wu. New convolutional codes - part III. *IEEE Trans. Comm.*, COM-24(9):946–955, 1976.
[479] W.W. Wu, D. Haccoun, R. Peile, and Y. Hirata. Coding for satellite communication. *IEEE J. Sel. Areas in Comm.*, SAC-5(4):724–748, 1987.
[480] A.D. Wyner and R.B. Ash. Analysis of recurrent codes. *IEEE Trans. Inf. Theory*, IT-9:143–156, 1963.
[481] H. Yamamoto and K. Itoh. Viterbi decoding algorithm for convolutional codes with repeat request. *IEEE Trans. Inf. Theory*, IT-26(5):540–547, 1980.
[482] Q. Yang and V.K. Bhargava. Optimum coding design for type-I hybrid ARQ error control schemes. *Elec. Let.*, 25:1595–1596, 1989.
[483] Q. Yang and V.K. Bhargava. Reducing time delay ARQ protocols by simple error-correcting codes. *Elec. Let.*, 26:1447–1448, 1990.
[484] Q. Yang and V.K. Bhargava. Delay and coding gain analysis of a truncated type-II hybrid ARQ protocol. *IEEE Trans. Veh. Tech.*, 42(1):22–32, 1993.
[485] H. Yashima, I. Sasase, and S. Mori. A new type of Viterbi decoding with path reduction. In *Proc. IEEE GLOBECOM Conf.*, pages 1714–1718, 1989.
[486] Y. Yasuda, Y. Hirata, K. Nakamura, and S. Otani. Development of variable-rate Viterbi decoder and its performance characteristics. In *Proc. Sixth Intl. Conf. Dig. Sat. Comm.*, pages XII–24 - XII–31, 1983.
[487] Y. Yasuda, K. Kashiki, and Y. Hirata. High-rate punctured convolutional codes for soft decision Viterbi decoding. *IEEE Trans. Comm.*, COM-32(3):315–319, 1984.
[488] Ø. Ytrehus. Constructions and proofs of nonexistance of some convolutional codes. Technical Report 60, Dept. of Informatics, U. Bergen, Bergen, Norway, 1992.
[489] Ø. Ytrehus. A note on high rate binary convolutional codes. Technical Report 68, Dept. of Informatics, U. Bergen, Bergen, Norway, 1992.
[490] F. M. Yuan. Minimal memory inverses of linear sequential circuits. *IEEE Trans. Comp.*, c-23(11):1155–1163, 1974.
[491] G.W. Zeoli. Coupled decoding of block-convolutional concatenated codes. *IEEE Trans. Comm.*, pages 219–226, 1973.
[492] L. Zhang and K.W. Sarkies. Modeling a virtual path and its application for forward error recovery coding schemes in ATM networks. In *Proceedings of SICON '91*, Singapore, 1991.
[493] H. Zhou and R.H. Deng. A hybrid ARQ scheme with diversity combining for land mobile radio. In *Proc. IEEE Veh. Tech. Conf.*, pages 902–905, 1992.
[494] K.S. Zigangirov. Some sequential decoding procedures. *Prob. Info. Transm.*, 2:13–25, 1966.
[495] K.S. Zigangirov. New asymptotic lower bound on the free distance for time-constant convolutional codes. *Prob. Info. Transm.*, 22:104–111, 1986.
[496] K.Sh. Zigangirov. On the error probability of sequential decoding on the bsc. *IEEE Trans. Inf. Theory*, IT-18(1):199–202, 1972.

# INDEX

**D**



**E**



**F**

## O



## P



## R



## S



## T



## U



## V

GPSR Compliance

*The European Union's (EU) General Product Safety Regulation (GPSR) is a set of rules that requires consumer products to be safe and our obligations to ensure this.*

*If you have any concerns about our products, you can contact us on ProductSafety@springernature.com*

In case Publisher is established outside the EU, the EU authorized representative is:

Springer Nature Customer Service Center GmbH
Europaplatz 3
69115 Heidelberg, Germany

**Batch number: 10755292**

Printed by Printforce, the Netherlands